KT-229-586

Contents

♈

Acknowledgements

I would like to thank staff and students of the Department of Geography at Lancaster University for providing the stimulus to write this book. In particular, I would like to thank Nicky Shadbolt for preparing the maps and diagrams, and those colleagues who have supplied photographs.

For Rebecca and Ruth,
who will have to live with whatever we do
to the world's climate

Introduction

Climatic change is a topic of endless fascination. Day-to-day variations in weather conditions are a staple of conversation in many countries. In the developed world the increasing volume of foreign travel in recent years has given more and more people direct experience of climates different from those of their home areas. It is normal to compare present weather conditions with those of past years. On longer time-scales older people often (and generally unfavourably) contrast present-day weather conditions with those of their youth. There is also widespread realisation that climatic conditions in the more distant past were different from those of today. This is highlighted by evidence such as early photographs showing Alpine glaciers in much more advanced positions than they are at present, or seventeenth-century paintings of 'frost fairs' on London's frozen River Thames.

We are aware that past human societies, with less developed technologies, were more vulnerable than ourselves to short-term climatic disasters – such as droughts, floods and storms – leading to harvest failure and famine. The survival of rock paintings in the middle of the Sahara, and large abandoned villages in the arid south-west USA, or speculation regarding the fate of the Norse Greenland colony reminds us that past cultures could be seriously affected by longer-term climatic shifts as well as year-to-year disasters (Wigley et al., 1981). We know that there have been periods in the past far colder than today, when ice sheets covered extensive areas of the northern land masses, as well as warmer eras such as those when the dinosaurs flourished.

In an increasingly overpopulated world the impact of climatic disasters is brought into our homes by the media. Natural disasters are often exacerbated by political instability or economic mismanagement. Such disasters frequently affect underdeveloped countries, but recent hurricanes in Britain and droughts in the USA demonstrate that climate can also have a substantial effect on the developed world.

The possibility of future climatic changes having a serious impact on mankind used to be the sole preserve of science fiction writers who have entertained, and sometimes scared, us with their tales of returning glacial conditions, global warming, and a rising sea level. Within the last decade, however, climatic change has stepped outside the realm of science fiction to become a real and worrying possibility. In recent years there has been an explosion of awareness and concern over environmental issues among the general public within the industrialised world. Some worries are local, such as the landscape changes caused by the construction of a new motorway. Other issues are significant on a regional scale – the problem of acid rain, for example. Some issues are truly global and, of these, climatic change is probably the most important in that it is likely to affect everyone to a certain extent in the future. Coming to terms with the problem by trying to adapt to climatic change or attempting to reduce its impact by cutting greenhouse gas emissions also requires a truly global approach.

In the late 1980s and early 1990s the terms 'greenhouse effect' and 'global warming' came into everyday use and there was a tremendous upsurge of interest in climatic change. At a time when green issues were receiving unprecedented attention, worry about global warming was fuelled by the media in a spate of popular books, articles, television documentaries and news items. It was hard to open any current affairs magazine without finding a reference to global warming and its possible consequences. Such media often made dire predictions about floods, droughts and famines.

We have had detailed knowledge of the ability of carbon dioxide and water vapour to absorb heat since the mid-nineteenth century. In 1895 a Swedish scientist, Svante Arrhenius, made the first attempts at linking variations in the average surface temperature of the Earth with changes in the atmosphere's content of carbon dioxide, as a possible mechanism for explaining the occurrence of ice ages. In 1938 an English scientist, G. S. Callendar, was the first to suggest that the addition of carbon dioxide to the atmosphere by human activity might be a factor behind the warmer conditions which were occurring in the early decades of the twentieth century. The International Geophysical Year in 1957 initiated the global monitoring of atmospheric carbon dioxide concentrations which provided the first clear evidence of a steady build-up of greenhouse gases. By the 1960s, possible links between the emission of greenhouse gases and global climatic trends were starting to impinge on the consciousness of people outside the scientific community. However, the trend of temperatures in the northern hemisphere during the 1960s and early 1970s suggested that cooling rather than warming might be setting in. This led to concern that industrial and agricultural pollutants in the atmosphere might be blocking out solar radiation, encouraging a possible downward slide of temperatures into a new ice age.

Nevertheless, the early 1970s also saw the first experiments in three-dimensional computer modelling of climatic change, crude though these may seem now, and the first catastrophic predictions that the West Antarctic ice sheet might disintegrate, as a result of rising sea levels caused by higher temperatures, raising world sea levels further by several metres.

In the early 1980s the rising concentration of various greenhouse gases in the atmosphere and their possible effects on climate, at both a regional and global level, began to attract increasing attention from scientists in a range of disciplines and concern began to be expressed about the potential human impacts of such changes. The topic had yet to attract widespread popular interest and was only just beginning to register in political circles. Interest in climatic change was reflected in a growing number of major international conferences and working parties and associations such as the World Meteorological Organisation, the International Council of Scientific Unions, and the Scientific Committee on Problems of the Environment. These established that increases in greenhouse gases due to human activity were potentially a major international problem and their press releases attracted growing media coverage.

In 1985, delegates at a conference in Villach, Austria issued a statement saying that mean global temperatures were likely to rise by between 1.5°C and 4.5°C over the next century, an estimate which is still widely accepted. The high level of consensus among scientists attending this conference was important in raising public consciousness of the global warming issue. In 1985, scientists working for the British Antarctic Survey published a paper confirming the existence of a hole in the ozone layer in the stratosphere above Antarctica. The problem of ozone depletion is linked to global warming by the fact that chlorofluorocarbons, or CFCs, the man-made gases responsible for much of the ozone depletion, were also powerful greenhouse gases. The discovery of the Antarctic ozone hole showed that, whatever the problems inherent in establishing the reality of global warming, man was clearly capable of inadvertently causing serious modifications to the atmosphere. The Montreal Protocol of 1987, an international agreement to try and halt ozone depletion by cutting the production of man-made CFCs, was also the first political action aimed, indirectly, at cutting emissions of a greenhouse gas.

The 1980s also saw a marked change in average global surface temperatures. The decade contained six of the warmest years since widespread instrumental records began about 150 years ago. In addition to this sudden jump in temperature, an increasing range of weather anomalies created a widespread feeling that weather patterns were changing, leading to increased interest in the possibility of man-induced global warming. More cynically, it has been suggested that, with the collapse of Communism in Europe and the USSR, the threat of nuclear war receded and people were looking for another global doomsday scenario with which to scare themselves, especially as the end of the millennium, a traditional time for dire predictions, was fast approaching.

The year when global warming began to attract widespread public attention and became a political issue has been pinpointed as 1988. In the USA heat and drought hit the Midwest and the Great Plains. Dry conditions encouraged forest fires, including the one that devastated Yellowstone National Park. The Mississippi fell to unprecedented low levels, disrupting river navigation and leaving thousands of barges stranded.

Also in 1988, the World Meteorological Organisation and the UN Environment Programme established the Intergovernmental Panel on Climate Change (IPCC). This group of nearly 400 scientists was given the task of assessing and summarising current knowledge on climatic change, particularly the likelihood of global warming. Two other working parties were concerned with the potential impacts of climatic change and possible responses to these impacts. In 1989, at the Toronto Conference on the Changing Atmosphere, with the benefit of improved computer models, there was a call to initiate practical measures to cut emissions of carbon dioxide. There was general agreement at the conference that the effects of global warming were likely to prove second only to global nuclear war if man did not start cutting back greenhouse gas emissions on a substantial scale. Many politicians were reluctant to take any action until the IPCC report appeared even though it was only intended to summarise existing knowledge rather than present new findings.

The initial IPCC scientific report appeared in 1990 (Houghton et al., 1990) with a follow-up in 1992 (Houghton et al., 1992). Although it has been criticised for brushing aside controversies and failing to emphasise uncertainties, it has become an important yardstick against which new research findings can be assessed. The report on the impacts of climatic change suffered from appearing at the same time, a fact which prevented its authors from using all the data generated by the first working party. The report of the third working party, on responses to climatic change, was even more speculative and the least successful of the three (Henderson-Sellers, 1993). The scientific assessment called for a start on discussions about political action. In 1992, the second UN Conference on the Environment and Development was held in Rio de Janeiro. One of the major results was the signing of a Framework Convention on Climate Change which represented a move towards concerted action on the reduction of greenhouse gas emissions (Chapter 7). The results were disappointing in many ways, due in part to entrenched positions and tensions between developed and less developed nations over resources and their exploitation. Consequently, agreements on measures to try and reduce emissions of greenhouse gases failed to specify targets for each country, producing a weaker result than had originally been hoped for.

During the 1990s, notably at the Rio summit in 1992, global warming appeared increasingly on political agendas, though real progress has been slow. The Montreal Protocol of 1987 showed what could be achieved by concerted political action when a direct and immediate threat to the atmosphere could be identified. Ozone depletion is, however, a process which can already be accurately measured and can be fairly confidently forecast for the future in relation to various levels of CFC production. The problem with global warming due to the man-enhanced greenhouse effect is that there is a lack of agreement about what is happening at present and much uncertainty about the future. While an increasing number of scientists believe that global warming will occur, and many consider that it is indeed already happening, there is not a complete consensus. Few scientists have been so bold as to commit themselves in the way that Jim Hansen of NASA did when, in 1988, he told a US Senate Energy Committee that he was 99 per cent certain that global warming was already in progress (Schneider, 1990).

One problem with the presentation of ideas on future climatic change is that much media coverage of issues relating to global warming has been alarmist and sensationalist, taking extreme scenarios and presenting them as near-certainties with little discussion of the probabilities of their occurring or the possible time-scales involved. Media reportage of scientific debate over global warming has often overlooked areas of consensus and focused on extreme positions in a way more appropriate to political debates. This has sometimes given the mistaken impression that major disagreements exist within the scientific community and that scientists do not have a clear view of what is happening. Additionally, because projections for future climatic change due to greenhouse gas increases take us well beyond the range of observed experience of climate within historic times, it is inevitable that many people find them hard to accept and are extremely sceptical.

Some reports, for example, have presented a catastrophic picture of a sea level rise of 5–6 m occurring within a century, when the scientific consensus was between 0.5–1 m (Chapter 5). Redrawn maps of the British Isles with Cambridge on the coast, Blackpool as an island and most of central London under water made good newspaper headlines but were so sensationalist that they damaged the credibility of scientific research. More recently a survey of global climatic change, as a contemporary issue in geography for A-Level students, opened with a map of Europe showing the areas which would be flooded with a sea level rise of 60 metres without any discussion of the likelihood of this occurring (Dawson, 1992).

The media are geared to short-term news stories and find it hard to sustain interest in long-term topics such as climatic change. More recently there has been a backlash with articles debunking what has sometimes been called the 'Greenhouse Effect Industry'. Scientists, too, have gone into print accusing colleagues like Jim Hansen and Stephen Schneider of being alarmist and misleading the public. Articles have claimed that the scale of global warming has been greatly overstated and that scientists have a vested interest in putting forward dire scenarios in order to attract further research funding. It has also been claimed that the data and methods they have used incorporate serious flaws which have distorted their conclusions.

A recent report in *The Sunday Times* was headlined 'Met Office Says Global Warming Is Just a Myth'. However, the report itself indicated that the Meteorological Office's study did not suggest that global warming would not occur, but that the rate of warming would be slower than suggested in the IPCC report; about 0.15°C per decade instead of 0.3°C. What the news item failed to indicate was that warming, even at this reduced rate, was still twice as fast as that which had occurred during the past century, and much faster than natural climatic shifts within recent historical times. The article indicated that the report was based on computer simulations without mentioning the areas of uncertainty involved in such techniques (Chapter 4). It also failed to point out that the amount of warming might be much greater than the global mean figure in some regions, though it did quote Professor Martin Parry as emphasising that agriculture, especially in drier areas, could still be at risk from changes of this scale. The uncritical reader might have considered that the report buried the global warming issue for good. However, less than two weeks later, television news carried a feature on the British Antarctic Survey's discovery that a dramatic spread of vegetation was occurring in the Antarctic Peninsula as a result of recent rises in average temperatures. One of the aims of this book is to help the reader to interpret such reports more effectively by explaining the background issues, and to show why there is not necessarily any contradiction between two such seemingly conflicting messages.

Recent revised views, forecasting less dramatic climatic change in the future, have resulted in part from research which has highlighted feedback mechanisms within the Earth's environmental systems which may work to damp down the scale of global warming and delay its onset. This was a prominent feature of the IPCC update report of 1992, causing estimates of the scale of global warming within the next few decades to be revised downwards (Houghton et al., 1992). On the other hand, (as discussed in Chapter 3), the operation of many feedback processes linked to climate are still poorly understood, even in some cases whether they may enhance or offset a warming trend caused by the build-up of greenhouse gases. So there may still be some unwelcome surprises in store as well as good news.

Reduced media coverage of climatic change issues may indicate that, after the big burst of green enthusiasm in the late 1980s and early 1990s, people are less concerned about global warming. Fears have been expressed by scientists that the case for global warming may have been stated too forcefully and that, if nothing serious appears to happen to the world's climate immediately, particularly if there is a short-term run of cooler years after the warm 1980s and early 1990s, public interest will wane and credibility will be lost.

Nevertheless, the terms 'greenhouse effect' and 'global warming' have become part of everyday language in the developed world, even if surveys of politicians have sometimes revealed widespread ignorance of what the greenhouse effect actually is and how it works. Opinion polls have shown that most people in Europe and North America have heard of the greenhouse effect while a significant proportion are worried about it and believe that their governments should be doing more to tackle the problem of global warming. People have a natural fascination for 'what if' questions: if global warming occurs roughly as forecast, what effects will it have on sea levels, on agriculture, on ecosystems, on everyday life? How quickly is significant climatic change likely to occur? What measures can be taken to slow its arrival or halt its progress?

Global warming is linked to many other environmental issues at both a global and regional level, including ozone depletion, deforestation, desertification and the occurrence of short-term weather anomalies bringing floods and droughts. The study of climatic change and its impacts is an interdisciplinary one involving research on past climates as well as those of the present and the future. Details of the data, techniques and concepts used by scientists working in this area are complex and may not be easily grasped by people who do not have a strong background in mathematics and physics. New research findings appear mainly in academic journals, monographs and edited collections of papers. These are not readily accessible outside university libraries and, being written for specialists, they make no concessions to the general reader. New research findings may be reported by the media and eventually find their way into paperbacks aimed at the general reader. Some of these, however, are simplistic and biased in their approach.

This book is aimed at students on geography and environmental studies courses in universities and colleges. It provides an introduction to the topic which discusses the basic concepts involved in the study of climatic change and its impacts, and is designed to provide a lead-in to more specialist literature. It assumes that readers have some basic knowledge of atmospheric processes, meteorology and climatology but do not necessarily have a strong scientific background. Emphasis is placed on how our understanding of the atmosphere has developed in recent years, but the limits of current knowledge are also stressed. Understanding climatic change is made more difficult because it results not only from many complex sets of interactions within the atmosphere, but between the atmosphere and other environmental systems. Many of these linkages, as we will see, are not well understood and there are still huge areas of uncertainty.

Nobody has yet been able to prove conclusively that global warming due to man's activities is actually in progress now or will definitely occur in the future. The problem is that climate, like any of the Earth's environmental systems, is subject to a considerable amount of natural variation from which it is difficult to separate any components due specifically to human action. Although many scientists believe that global warming is already under way, it is widely accepted that another two decades, perhaps more, of further research, may be necessary before conclusive proof is available. The dilemma is that, by the time such evidence is forthcoming, it may be too late to take effective action to counter global warming. The background to understanding this dilemma forms the core of this book.

Because it is essential to understand the range of variability of the Earth's climate on different time-scales before we can fully appreciate how it operates at present and how it may change in the future, this book begins with two chapters on past climates and how they can be reconstructed. A chapter on the greenhouse effect and greenhouse gases looks at how man has changed the composition of the atmosphere and the ways in which this may affect climate. The difficulties involved in predicting future climates, with the strengths and limitations of the techniques which climatologists use, are then considered, highlighting current uncertainties. Attention is then focused on the possible impacts of future climatic changes. Finally, the political and economic dimensions of climatic change are discussed, including the measures which are currently being taken to reduce emissions and possible future action. Although this book aims to explain the concepts involved in the study of climatic change in as straightforward a way as possible, the use of technical terms cannot be avoided. These are explained whenever they first appear in the text but they are also gathered together in a glossary at the end of the book.

1

Reconstructing Past Climates

It may seem strange to begin a study of climatic change which focuses on the future with an excursion into the past, not just to recent centuries but to the last ice age and beyond. However, an axiom of the well-known climatic historian Professor Hubert Lamb (1982) is that in order to understand present-day climate, and how it might vary in the future, it is first necessary to appreciate how and why climate has fluctuated in the past. Information about past climates provides a basis for testing hypotheses about the causes of climatic change. By reconstructing past climates on a range of time-scales we can gauge the degree of natural variability in the climate system. The '**forcing factors**' (changes imposed on the climate system which alter its radiative balance) which have operated in the past to produce climatic changes will continue to affect climate in the future. Any changes resulting from the man-enhanced greenhouse effect will be superimposed upon, and will interact with, these natural fluctuations.

Reconstructing past climates helps to place contemporary climates in perspective, making it easier to identify trends and periodicities. Unless we can understand the natural variability of climate it will be hard to isolate those influences deriving from human activity. Climates from the recent and more distant past provide yardsticks against which we can assess the scale of possible future climatic changes, especially those resulting from greenhouse gas increases.

Having stressed this, it is sobering to reflect that the upper end of the range of temperature increases which have been predicted to occur by AD2100 – 5.5°C above present averages – would make global temperatures higher than they have been at any time during the last three million years, bringing them close to what they are thought to have been in the Cretaceous period over 65 million years ago. A more modest increase of 3°C would still produce warmer conditions than at any time in the last 125,000 years, while a rise of only 2°C would take us beyond anything that has been experienced since the end of the last glaciation.

But how do we know what climate was like in the past? Instrumental weather records are available for the last 200 or so years providing a precise, quantitative (though far from complete), picture of climatic variation during the period in which human interference with the composition of the atmosphere has become increasingly evident. Going further back we have less and less information about past climatic conditions. However, qualitative weather observations and many indirect or **proxy** indicators allow more general reconstructions of climate in some parts of the world back to medieval times (Flohn, 1985). Beyond this, and more recently for those parts of the world where written records begin late, a wide range of scientific techniques can be applied to different kinds of environmental evidence to provide a picture of climatic variation. The study of palaeoclimates is a complex, inter-disciplinary field and it is impossible here to do more than briefly review some of the most important methods of climate reconstruction which are currently available, starting with those relating to more recent times before moving on to deal with those which concern the more distant past. Having examined these techniques, we can then consider the evidence they provide about past climatic variations, beginning on a time-scale of hundreds of thousands of years and moving towards the twentieth century with an increasingly sharp focus.

INSTRUMENTAL RECORDS

The thermometer, rain gauge and barometer were invented in the seventeenth century but very few surviving sets of measurements are as old as this. Only for a handful of places do quantitative meteorological data go back more than 200 years. The longest continuous run of temperature records, beginning in 1659, was pieced together by Professor Gordon Manley (1974) from various sets of data relating to Central England (Fig. 1.1). His monthly mean surface temperatures and longer-term averages were based, in the earlier part of the series, on amalgamations of short runs of observations from different locations, so that accuracy for the seventeenth and early eighteenth centuries is a little shaky. The earliest of the seventeenth-century values may only be reliable to within 1°C and those for the first half of the eighteenth century to within 0.2°C. Nevertheless the Central England temperature series extends back to the coolest phase of the Little Ice Age (Chapter 2) in western Europe and provides a useful yardstick against which other types of climatic data can be measured.

Meteorological observations start for The Netherlands and Berlin in the early years of the eighteenth century (Bradley and Jones, 1992). By the end of the eighteenth century, stations were sufficiently numerous in

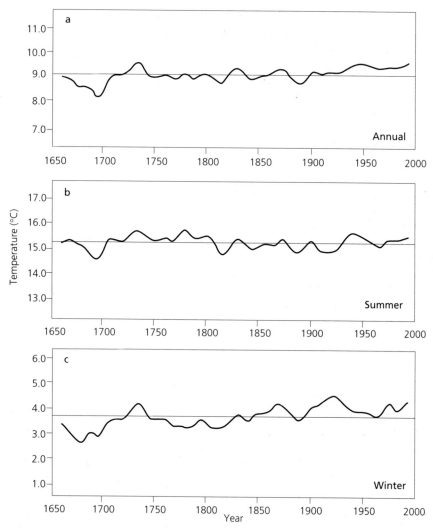

Fig. 1.1 Professor G. Manley's temperature series for central England, the
longest available run of instrumental records. (After Roberts, 1994)

Europe, and instruments sufficiently standardised, to allow the construction of daily weather maps from the mid-1780s onwards (Kingston,
1981). Geographical coverage nevertheless remained thin for many parts
of eastern Europe and the Mediterranean well into the nineteenth
century. For many areas of the world instrumental records go back less
than a century. Before this observations were mainly confined to Europe
and North America: they began in the mid-nineteenth century for most
parts of South America, Asia and Australasia. The longest temperature

series for the southern hemisphere, from Rio de Janeiro, only starts in 1832. By 1900, however, only a few areas outside the polar regions and the interiors of Africa, Asia and South America were still without coverage. Data are still sparse for polar areas; records did not start for parts of the Canadian Arctic and Antarctica until the 1940s and 1950s respectively.

The interpretation of past meteorological records is far from straightforward. Even today the coverage of weather stations is heavily biased towards the developed world and data on atmospheric conditions are far more abundant over land than over oceans. Changes in the types of instruments used mean that figures within any time series, or from different stations, may not be strictly comparable without some adjustment. Changes in the siting of instruments at particular weather stations and alterations in the environment immediately around them can cause variations which have nothing to do with climate. Temperature records can show sudden jumps or falls due to changes in the amount of heating in the buildings to the external walls of which thermometers were fixed.

As towns and cities grow they generate their own distinctive climate, warmer than the surrounding countryside, due to the capacity of building materials to absorb radiation more readily than vegetation. Many city-centre temperature records thus show a gradual warming trend which is purely local in origin. This has sometimes been demonstrated when a weather station has been moved from a central to a peripheral location within an urban area and mean temperatures have suddenly dropped. Some critics have indeed claimed that the twentieth-century global warming trend (Chapter 2) is no more than the effect of the build-up of local urban heating on temperature records. Such influences can, however, be allowed for. They can be isolated by comparing temperature trends for urban meteorological stations with those in nearby rural locations. Local anomalies can then be removed from the data set (Jones et al., 1989). Estimates of the impact of urban warming on mean global temperatures over the past century are thought to be only in the region of 0.01–0.05°C, much smaller than the overall temperature rise. Other kinds of data also need careful handling. Changes in sea surface temperatures, for instance, are also likely to have been partly due to differences in the types of buckets used for hauling up samples; wooden, metal or canvas ones were normal at different periods and variations in the degree to which they absorbed heat from samples of sea water affected the temperatures recorded.

The term 'mean global surface temperature' is used throughout this book in relation to past, present and future climate. It is worth pausing for a moment to consider some of the complexities involved in calculating such figures and the problems of interpreting them. First, as we have seen, until recently the geographical coverage of land-based meteorological stations was very uneven. Coverage is still biased towards North

America and Europe. Even within a well-recorded country like Britain there is a lack of data from upland areas. Different methods are used to calculate mean daily temperatures for stations in various countries and often for the same station through time, so that corrections have to be applied to make data sets comparable. Vast quantities of data have to be amalgamated to produce global mean figures and the hope is that any small-scale errors will be cancelled out rather than be cumulative. This makes mean global surface temperature, despite its widespread use, more of a blunt instrument than a sharp probe as far as examining climatic variation is concerned.

HISTORICAL AND ARCHAEOLOGICAL DATA

Instrumental records only cover a short time-span of climatic history, one where there has been a steady trend towards warming, and a much longer perspective is needed to identify and understand the full range of climatic variation which has occurred. For many parts of the world, qualitative records of weather conditions and weather-related phenomena – such as droughts, floods, the freezing of rivers and lakes, the flowering of trees and the ripening of grapes – provide information about past climates which is less precise than instrumental records but is more abundant and sometimes extends several centuries further back in time (Ingram et al., 1981). Ships' logs, for example, contain a good deal of information on weather conditions (Oliver and Kingston, 1970). Medieval manorial records are a useful source of information on weather events such as droughts, severe snowfalls and storms, as well as providing information on the impact of such events on society (Titow, 1960 and Dury, 1984).

Historical data, however, have to be interpreted with care. It is necessary to understand the background of diarists and observers. Early descriptions of weather conditions in North America should be treated with some caution if the writers had not been in an area for long enough to gain an impression of what constituted 'normal' conditions (Bradley and Jones, 1992). A winter described as 'hard' by a seventeenth-century diarist like Pepys was not necessarily comparable with one described in similar terms by a medieval monastic chronicler. The frost fairs on the River Thames in Pepys' day certainly reflected winter conditions colder than those experienced in recent times but they may also relate in part to the 'ponding back' of fresh water by old London Bridge, making the river freeze more easily than it does today.

A knowledge of pressure and circulation patterns has allowed climatic historians to take individual stray weather observations and combine them to produce seasonal pictures of conditions across Europe as far back as the fifteenth century (Lamb, 1982). The French historian Le Roy Ladurie (1971) established trends in post-medieval climate in the Alps by reconstructing glacier fluctuations from maps and documentary sources. A similar approach by Jean Grove (1985, 1988), using Norwegian tax records, has established a rather different chronology of advances and retreats of glaciers and climatic changes in southern Scandinavia between the sixteenth and nineteenth centuries. Le Roy Ladurie (1980) also pioneered the use of local records dating the start of the grape harvest in France and adjacent countries. Variations in the date at which the harvest commenced provided a useful indication of conditions during the previous spring and summer going back to late medieval times.

Archaeological data can be used to examine the relationship between climatic changes and fluctuations in settlement and cultivation limits, though care must be taken not to apply a simple 'cause and effect' approach (McGhee, 1981 and Neuman, 1993). Parry (1977), studying the Lammermuir Hills in south-east Scotland, was able to show that, during the twelfth and thirteenth centuries, with warmer and drier conditions than today, settlement and cultivation expanded over all but the highest part of this rolling plateau area, with cultivation occurring at altitudes where crops of oats would not ripen in almost any year at the present time. He also linked the progressive abandonment of settlements and their field systems in this area from the fourteenth to the eighteenth century to the steady downward shift in viable cultivation limits as conditions became progressively colder and wetter (Figs. 1.2, 1.3). In highlighting the changes in cultivation limits, Parry was careful to stress that climatic change was probably only a background influence which may have helped to tip the balance in favour of abandonment when other short-term economic, social and political factors were unfavourable.

These examples relate to the more densely populated areas of western and central Europe, one of the best documented parts of the world in the last few centuries. For some more remote areas where climatic conditions are particularly sensitive we are fortunate in having detailed written observations predating instrumental records. For example, information on the extent of summer sea ice around Iceland goes back to the Norse settlement in the ninth century (Ogilvie, 1984). For Arctic Canada, records of the Hudson's Bay Company provide a wealth of data on past weather and sea ice conditions from the mid-eighteenth century (Ball, 1992 and Catchpole, 1992). The Company's ships followed regular sailing routes from England to the Hudson Strait at the same time each year in order to fit in with the short ice-free season. The logbooks of the vessels, which were kept in a standardised format from 1751, provide details of sea ice conditions. By comparing modern year-to-year sea ice variations with climatic data it is possible to use the logbook entries to make deductions about past circulation patterns and temperature conditions. Records from the Company's trading posts also contain a wealth of weather-related

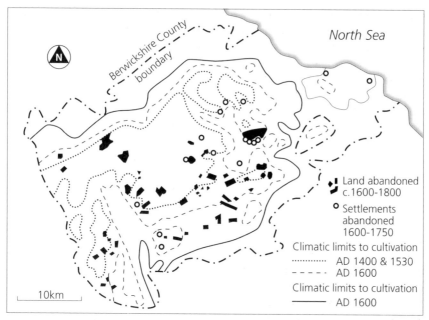

Fig. 1.2 Abandoned farmland and lowered climatic limits to cultivation in the Lammermuir Hills, south-east Scotland, 1600–1750. (After Parry, 1977)

Fig. 1.3 Abandoned cultivation ridges at high altitude in southern Scotland: a casualty of the Little Ice Age? (Photo: I. Whyte)

information – such as the annual dates of freezing and thawing of rivers like the Red River at Winnipeg, and the date of the arrival of migratory geese in spring – together with direct weather observations such as the onset of spring thaws and the first autumn snowfalls (Ball, 1983, Bradley and Jones, 1992, Catchpole and Faurer, 1983 and Rannie, 1983). Graphical representation of such data allows short-term warm and cold spells to be identified and linked to the position of the Arctic Front. On a longer time-scale, the start of the modern warm phase in the late nineteenth century is also readily detectable.

ENVIRONMENTAL INDICATORS

Even historical records do not take us back very far though. Fortunately a wide range of natural phenomena are climate-dependent and become sealed into stratified deposits containing built-in proxy measures of past climates (Bradley, 1985). A wide range of proxy data is now available relating to environments in different parts of the world. Each technique of climatic reconstruction has its own strengths and limitations. The further one goes into the past, the more broad-brush is the picture of climate they provide.

An important aspect of such indicators is the quality of their time resolution. Sources which provide data on a seasonal or annual basis, such as tree rings and ice cores, allow climatic fluctuations to be dated accurately. However, such information usually only extends back a few thousand years. Ocean sediment cores on the other hand provide proxy information on climate over hundreds of thousands of years but with a much more blurred time resolution, often of many centuries.

Some proxy records, such as ice and ocean sediment cores, are continuous. Others, like moraines marking the extent of glacier advances or shorelines indicating former high lake levels, provide discontinuous information. Proxy data respond to climatic changes at different speeds. Records of fossil beetles show that these insects responded rapidly to temperature changes while variations in fossil pollen demonstrate that plants took far longer to adjust to the same changing conditions.

Plant and animal remains

Fossil pollen grains, which can be identified to the level of the family and, for many trees, the genus, preserve well in peat bogs, buried soils and lake sediments. If cores are taken from such deposits and the pollen assemblages studied for samples at different depths, variations in the vegetation cover over time can be identified. Inferences can then be made regarding the environmental changes, including climatic fluctuations, that caused the vegetation to alter. Pollen analysis was developed in Sweden and Britain in the early twentieth century and has been widely used in reconstructing postglacial environments, at first within Europe and more recently elsewhere in the world (Lowe and Walker, 1984, Birks and Birks, 1980 and Birks, 1981). The time resolution of the environmental changes preserved in the pollen record is fairly coarse as the response of vegetation to climatic changes generally involved a substantial time lag. Nevertheless, the analysis of pollen variations from sites in Britain and elsewhere in western Europe provided the first clear chronology of climatic changes in the early part of the **Holocene** (postglacial) period. However, it was soon recognised that much of the variation in pollen assemblages could be attributed to purely local environmental changes, such as drainage conditions, rather than the direct effects of climate (Barber, 1985). As human population increased, especially with the spread of agriculture, changes in the pollen record came increasingly to reflect local human interference in the vegetation cover rather than background changes in climate. In addition, pollen macro-remains of plants, such as trees buried at the base of growing peat, provide information on altitudinal and latitudinal vegetation limits and have been used in calculating summer temperatures at the period of maximum postglacial warmth.

Animal remains can also provide clues to past climates. The hard parts of Coleoptera (beetles) are often abundant and well preserved in peat and sedimentary deposits. Fossil beetle remains can frequently be identified at species level. From studying their modern counterparts it is known that different species have marked preferences for particular environments and can often thrive only within set temperature ranges. Their ability to respond quickly to temperature changes and to colonise suitable new areas makes them a much faster-moving indicator of climate than plants (Coope, 1977). During the last interglacial period the assemblage of beetles inhabiting lowland England was comparable to that found in southern Europe today, suggesting that summer temperatures were around 3°C higher than they are now. Because of their quick reactions, studies of changes in beetle assemblages during the last 50,000 years have allowed the identification of warm periods which were too short-lived to be registered by changes in vegetation. For example, about 43,000 years ago temperatures for perhaps 1,000 years or so were 1–2°C above those of today. Central England nevertheless remained devoid of trees which did not have sufficient time to migrate in before the return of colder conditions and this warm phase does not register in the pollen record. A similar series of events occurred during the Windermere **Interstadial** (a brief warm phase) around 13,000–12,500 years ago. By the time the birch woods had moved into southern Britain and were beginning to spread, the peak of warmth had already passed and assemblages of beetles were changing with an increase in the proportion of less warmth-loving species.

Glacier fluctuations

Glacier fluctuations are the result of changes in the balance between the accumulation of snow which nourishes them and loss by melting or the calving of icebergs. A glacier advance reflects a positive mass balance due to climatic conditions which encourage accumulation over melting. Glacier

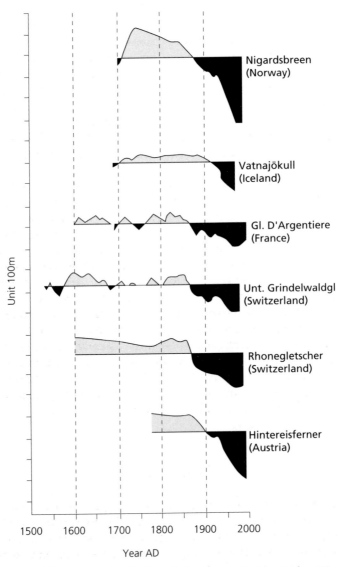

Fig. 1.4 Advance and retreat of selected European glaciers. (After Warrick and Oerlemans in Houghton et al. (eds), 1990)

fluctuations provide a good barometer of past climatic variations because they can amplify small changes in temperature and/or precipitation into perhaps several hundred metres of ice advance or retreat (Porter, 1981). The rise in mean global temperatures of about 0.5°C which has occurred in the last century has been accompanied by the worldwide retreat of glaciers and small ice caps (Lamb, 1982) (Fig. 1.4). The spectacular scale of this retreat can often be calculated by comparing present glacier limits with those recorded on old maps or early photographs.

Earlier periods of glacier advance in Europe can be established from documents and moraine limits. Moraines can be dated by a variety of techniques such as radiocarbon dating or lichenometry – measuring the growth rates of lichens on boulders incorporated in moraines to work out roughly when they first emerged from under the ice (Matthews, 1975 and Matthews and Shakesby, 1984). The morphological record of glacier variations is, inevitably, only a partial one because any ice advance tends to obliterate traces of previous movements of lesser magnitude. However, if all the glaciers within an upland region are considered, fragments of moraine from a whole series of earlier fluctuations can often be identified (Karlen, 1973).

The response of glaciers to climatic changes is not simple though (Bradley, 1985). Many combinations of climate can produce changes in their mass balance. Variations in precipitation as well as temperature have to be considered. Colder winters may not necessarily bring more snow while milder winters with a more vigorous atmospheric circulation may lead to greater snowfall and accumulation. Equally, glacier advances can be linked to cool, cloudy summers when melting is reduced. There is also a time lag between changes in mass balance and the position of a glacier front, while downwasting of ice can occur without any change in the position of glacier margins. Time lags are a function of the length of the glacier, its slope, thickness and the overall geometry of its basin. Some glaciers may respond to climatic changes within 25 years or less, others take much longer. Due to variations in response times, different glaciers within the same region may be advancing and retreating simultaneously. Some Arctic glaciers are still advancing as a response to cooler conditions at the end of the Little Ice Age.

Lake levels

Former lake levels provide a low-latitude counterpart to glacier moraines as indirect evidence of past climatic fluctuations (Street-Perrott and Harrison, 1985). Much climatic reconstruction focuses on past temperatures and the study of changes in lake hydrology is important in directing attention to variations in precipitation. In arid and semi-arid areas much of the surface drainage flows not into the oceans but into enclosed basins. In such basins changes in the hydrological balance can have dramatic effects on water storage. Lakes in such situations can act, effectively, as giant rain gauges. The size of some past lakes during **pluvial** (wet) phases is impressive. Lake Chad, which now has a surface area of only about 10,000 km^2, covered 300,000 km^2 at its postglacial maximum, though it was very shallow reaching a maximum depth of only 50 m. In the south west USA the Great Salt Lake is only a shrunken remnant of the much larger pluvial Lake Bonneville which covered 50,000 km^2 and was over 330 m deep.

In closed lake basins, high lake levels may reflect wetter phases or periods when evaporation was reduced. Low lake levels reflect arid spells or periods of increased evaporation. Whether variations in the levels of pluvial lakes reflect changes in precipitation, evaporation or both is a controversial topic but the question can be explored using hydrological balance models, while there is often enough evidence of past temperature conditions to provide an indication of which was the most important variable at work. Former high lake shorelines can be recognised by features such as abandoned cliffs, beaches, bars and deltas. Periods of low lake levels can be determined by analysing cores taken from lake floor sediments, with layers of evaporite marking periods when lakes dried up completely (Fig. 1.5. Street-Perrott, 1994).

Much early work on postglacial climate history focused on higher latitudes rather than on tropical areas. The evidence of lake level fluctuations has provided valuable indications of the climatic history of Africa.

Fig. 1.5 Salt crust on dried bed of Lake Magadi, Kenya. (Photo: Dr. P. Barker)

Radiocarbon dating of material in lake deposits has enabled former periods of high and low water to be dated, allowing correlation across the Sahel and East Africa (Street and Grove, 1979). It was originally believed that pluvial phases in semi-arid areas corresponded to periods of glacial advance in higher latitudes. The extensive pluvial lakes in the south-west USA did indeed reach their maximum extent at this time. Evidence from sub-Saharan Africa has shown, however, that the main pluvial phase was not contemporary with the last glacial maximum but with the postglacial climatic optimum (Chapter 2).

Studies of present water regimes and knowledge of past temperatures, allowing evaporation rates to be calculated, have enabled scientists to establish that, at the height of the postglacial pluvial phase, rainfall over East Africa was around 65 per cent higher than today and in the Sudan and Mauritania between twice and four times present precipitation levels (Roberts, 1989). Even in more recent times lake levels and precipitation have fluctuated significantly. For Lake Malawi, one of the world's largest lakes, sediment cores show that the lake dropped by 100 m between AD1450 and 1850 and that rainfall at this time may have been only 50–70 per cent of that of the twentieth century. The recent chronology established for East Africa differs from that of West Africa, where the level of Lake Chad was high in the sixteenth and seventeenth century and lower in the eighteenth and early nineteenth (Maley, 1977 and Nicholson, 1978)

Fig. 1.6 Drilling a sediment core in the bed of Lake Magadi, Kenya. (Photo: Dr. P. Barker)

demonstrating the regional contrasts which existed. Apart from showing that climate in sub-Saharan Africa has been as variable in recent centuries as in parts of Europe during the Little Ice Age, lake level data also show that the recent drought which has affected the area since the late 1960s is not unique and is more likely to be a natural fluctuation than a symptom of man-induced climatic change (Fig. 1.6).

Dendroclimatology

Dendroclimatology is based on the long-appreciated fact that trees in many parts of the world grow in an annual cycle, influenced by climatic variations, which leaves annual rings when the trunk is sectioned. The rings represent the alternation of large, thin-walled   cambium cells in spring and summer and more densely packed thick-walled cells later in the growing season. Trees occur over much of the world though in tropical climates the lack of seasonality may mean that clear rings do not form. Tree rings also have the advantage that they can be accurately dated, so that their time resolution is good. Wide rings indicate favourable environmental conditions, narrow rings adverse ones. In semi-arid areas the main influence on tree growth and ring width is the availability of moisture. On the cooler margins of temperate climate such as Scotland or Scandinavia the principal influence on tree growth is temperature (Fig. 1.7).

Dendroclimatology was pioneered early this century as a means of identifying precipitation variations in the arid south-west USA (Fritts, 1976 and Fritts et al., 1981). In this region living examples of Bristlecone Pine can be up to 5,000 years old, providing a chronology of precipitation variation with an annual time resolution reaching back halfway through the postglacial period. The use of wood from dead trees allowed the chronology to be pushed back even further to over 8,000 years. The technique has also been used in Europe and Russia, the Mediterranean, the Himalayas, China and Australasia (Bradley and Jones, 1992). It is

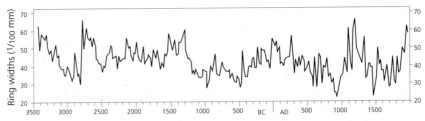

Fig. 1.7 Variations in ring widths (twenty-year moving averages) of Bristlecone Pine near the treeline in the White Mountains of California. In this location changes in ring width indicate variations in summer warmth. (After Lamb, 1982)

particularly useful for areas where instrumental records started comparatively recently. Work on trees from Irish peat bogs has produced a
chronology nearly as long as that for the Bristlecone Pine. Comparison of
tree ring width variations in recent times against modern climatic data
show which climate variables have most influence on tree growth in particular areas. This allows inferences about earlier climatic patterns to be
made from tree ring information alone.

Trees growing near the limits of survival of their species provide a better
record than those in more favourable locations in the middle of their
range. Tree ring width calibrations can be used to identify individual bad
years but the data are more useful if they are averaged out over longer
periods such as decades. The ring widths of individual trees contain variations due to the process of ageing, and changes in local site conditions.
Chronologies for particular locations are built up from sets of ring width
data from many individual trees. These are carefully chosen to reduce the
extent of purely local non-climatic influences. The technique has been
refined and extended throughout North America allowing the geographical distribution of past phases of drought and temperature variation to be
mapped and related to variations in atmospheric circulation (Fritts and
Lough, 1985).

Oxygen isotope data

One of the main sources of information on palaeoclimates on longer time-
scales is the oxygen isotope content of the water molecules of snow
crystals. An **isotope** is a variant of a normal element whose atoms have a
different atomic weight. For oxygen, atoms with the normal atomic weight
of 16 make up nearly 99.8 per cent of the total but 0.2 per cent are heavier
with an atomic weight of 18. These lighter and heavier atoms are present
in the atmosphere and in water. The proportion of ^{18}O atoms to ^{16}O is
around 1:500 on average but varies between 1:495 and 1:515 according
to temperature.

When oxygen is incorporated in a stratified deposit such as a layer of
snow on a glacier surface or in the calcium carbonate of a shell in ocean
sediments, the ratio between ^{18}O and ^{16}O atoms is frozen, providing a
record of the temperature conditions of air or water occurring at the time
that the deposit was laid down. Oxygen isotope data can be extracted
from various stratified deposits but the most useful for reconstructing past
climates have been cores from ice caps and ocean sediments.

The snow which accumulates each year on the world's ice caps and
glaciers is eventually compacted to form ice. Within the ice, however,
annual layers of accumulation can still be identified, sometimes to a
considerable depth. Variations in the oxygen isotope content, and other
features discussed below, provide a valuable climatic record. The isotope

ratio in a particular ice layer is a function of the temperature at the time that the original water vapour condensed.

Ice cores were first drilled into the Greenland ice cap in the 1960s and then in different parts of the Arctic and the Antarctic, choosing sites where the deformation of annual layers due to ice movement was minimal. New cores and more sophisticated methods of analysing them are providing more detailed information about climatic change on time-scales reaching back to the last interglacial period (Bolton, 1993). The Vostok core in the Antarctic has penetrated to ice which is over 200,000 years old but another 1 km or 500,000 years of ice has still to be drilled and analysed (Jousel et al., 1993). Fascinating as the details of climate suggested by these cores are, they nevertheless relate to areas remote from centres of population whose climates did not necessarily completely mirror global conditions. More recently it has been appreciated that, by careful choice of site to minimise dislocation caused by ice movement, useful data can be obtained from other areas; cores from small ice caps in Tibet and the Andes have provided the first details of climatic fluctuations in the tropics (Thompson, 1992).

It is generally assumed that oxygen isotope measurements from ice cores are a good proxy source of information about past temperatures. In fact other influences such as changes in patterns of storm tracks, affecting the origins of the moisture which produced the snow, can also influence ^{18}O levels. In addition the deeper layers may have originated up-glacier from the sampling point and have accumulated under slightly different climatic conditions. So it is not easy to translate ^{18}O variations directly into a record of temperature fluctuations though, for recent times, ^{18}O variations in snow and ice layers can be calibrated against instrumental records of temperature. Annual layers of accumulation in the ice can be detected from melt features and seasonal variations in the ^{18}O levels – higher in summer, lower in winter – but this is difficult to do beyond a certain depth. For the Camp Century core in Greenland it was possible to identify individual years going back around 10,000 years. The seasonal signal of the ^{18}O isotope is gradually smoothed out as the snow is compacted into firn and it may not be preserved at all where the annual rate of accumulation of new snow is low. This makes the identification and precise dating of annual layers of snow more difficult in areas such as the centre of the Antarctic where precipitation is light compared with Greenland.

As well as variations in oxygen isotope content, ice layers also contain other evidence relating to past climates. Differences in the thickness of the annual layer reflect changes in snowfall amounts. Layers of dust from volcanic eruptions are preserved and form useful chronological markers. The electrical conductivity of different layers of ice also varies. Ice formed during cold periods has a higher alkaline content and lower conductivity than ice from warmer periods. Electrical conductivity analysis provides a quicker and less expensive way of getting a broad-brush impression of changes throughout an ice core, allowing the identification of key periods of climatic change which can then be investigated in more detail by more labour-intensive methods, such as oxygen isotope analysis (Taylor et al., 1993).

The upper levels of snow in Greenland provide a sensitive barometer of other environmental changes due to man. Atmospheric lead levels, for instance, have risen around 200 times in recent years due to the use of leaded petrol in motor vehicle engines in the developed countries of the northern hemisphere. In the last 20 years, however, lead levels in Greenland snow have fallen dramatically, apparently due to the increasing use of unleaded petrol in North America from the 1970s (Rosman et al., 1993). More generally, variations in the dust content of ice layers provide indications of past patterns of storminess and changes in wind direction associated with different atmospheric circulation patterns. Past concentrations of greenhouse gases like carbon dioxide and methane can be calculated from bubbles of air trapped within the ice. The latest cores from Greenland, an area of high snow accumulation, allow climatic changes to be established on a time-scale of decades or less for the past 240,000 years, going back beyond the last interglacial period.

The study of oxygen isotope variations from ocean floor sediment cores provided the first long-term chronology of climatic change stretching back through the Quaternary era into the Tertiary (Bradley, 1985). Variations in the oxygen isotope composition of sea water are preserved in the shells and skeletons of marine creatures incorporated in sea-floor sediments as a result of the oxygen absorbed from sea water during the making of the shells. Lighter molecules of water containing ^{16}O atoms evaporate more easily than the ^{18}O ones. During glacial periods large quantities of ^{16}O were locked up in ice leaving the oceans relatively richer in ^{18}O atoms. The melting of ice sheets returned large volumes of ^{16}O enriched water to the oceans. The proportion of ^{18}O in marine shells depends partly on temperature conditions and partly on the isotopic content of sea water. It is believed, however, that the oxygen isotope variations primarily reflect changes in the volume of land ice rather than ocean temperatures. They thus provide a chronology of glacial advances and retreats which indirectly points to changes in climate (Fig. 1.8).

The ice volume signal is, however, a global one; it does not indicate the extent of ice in different areas. The time resolution of information from ocean sediments is poor due to mixing by bottom-dwelling organisms and turbidity currents. Often it is not much better than 1,000 years even when sedimentation rates are high. This blurring makes it difficult to compare cores from different areas and to pinpoint the timing of warmer and colder phases. Nevertheless, ocean sediments have the advantage over ice cores of being distributed over 70 per cent of the planet and they provide the best long-term source of information on palaeoclimates.

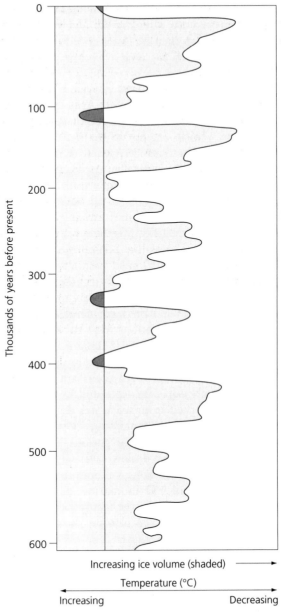

Fig. 1.8 Volume of the Earth's ice sheets and (indirectly) changes in temperature over the last 600,000 years, derived from isotope variations in sea floor sediments. The heavily shaded areas indicate periods which were as warm or warmer than the present. (After Lowe, 1993)

In recent years developments in the analysis of cores from ice caps and ocean sediments have provided clearer evidence of past climatic changes on time-scales of tens and hundreds of thousands of years. A major discovery has been that the Earth's climate, rather than changing slowly over long time periods, has often been very unstable, changing sharply and quickly. For instance, analysis of cores from the Greenland ice cap has shown that the transition to warmer conditions at the end of the last glaciation occurred swiftly; certainly within a century, possibly within 30 years. The end of the severe cold snap known as the Younger Dryas event about 11,500 years ago (see below) seems to have occurred within 20 years. A core from the Quelcaya ice cap in the Peruvian Andes shows that the post-medieval cool phase, widely referred to as the Little Ice Age, came within two or three years around AD1800 (Thompson, 1992).

This chapter has looked at some of the ways in which past climates can be reconstructed on time-scales ranging from annual to glacial/interglacial cycles. Taken together, the various sources of evidence examined provide an increasingly detailed picture of climatic variations. However, due to the complexity of the climate system and the forcing factors that have affected it, each new piece of evidence tends to throw up more questions than it answers. The next chapter looks at the scale and nature of the climatic changes that have occurred and some of the possible underlying influences which have caused them.

2

Climatic Changes in the Quaternary Era

The climatic changes considered in this chapter were brought about by various internal and external mechanisms. Fluctuations in the Earth's orbit have been important in affecting climate at time-scales of tens and hundreds of thousands of years, but even here, the explanation of climatic change is complex and many influences are likely to have combined to produce the patterns that may be detected from geological and biological records. Having finished the previous chapter with a consideration of the record of ocean sediments, it is useful to start this chapter with the long time-scale of the Pleistocene glaciations and work forward to more recent times at successively finer levels of detail.

THE GLACIAL/INTERGLACIAL SEQUENCE

The most notable aspect of climate during the past two million years has been the alternation between glacial and interglacial periods, between colder periods of major ice advance and warmer ones when the extent of ice was greatly reduced and temperatures were similar to those of today or even higher. The Pleistocene is remarkable not just for the alternation of warm and cold phases but for the scale and rapidity of the changes. At least 20 glacial/interglacial cycles are known to have occurred and within each longer cycle there may have been many more brief oscillations, often lasting only a thousand years or less. The long glacial phases were interrupted by short-lived periods of ice advance – **stadials** – and retreat – **interstadials**.

This was not the first of the Earth's ice ages; other phases of glaciation have been identified from the geological record. At least eight previous ice ages are known ranging through geological time from the Jurassic, around 150 million years ago, to the pre-Cambrian, 2,300 million years ago and beyond. Significantly, all these earlier ice ages lasted for longer – up to 50 million years – than the two million years of the Quaternary ice age (John, 1979). The Pleistocene ice ages are, however, far better recorded. Again and again temperatures dropped to a level which allowed the spread of sea ice in polar areas and a huge expansion of ice caps and glaciers in the northern hemisphere. At their maximum ice in Europe reached southern England with smaller ice caps in the Alps and Pyrenees. In North America ice limits pushed south of where the Great Lakes are today. Huge areas beyond the ice limits experienced an intensely cold or 'periglacial' climate similar to that of the modern Arctic. As more and more water became locked up in ice, sea levels fell by over 100 m.

In the first half of the Quaternary period the dominant cycle of glacial/interglacial conditions was about 40,000 years, with fluctuations in climate being less sharp as well as of shorter duration than in more recent times. Between 800,000 and 900,000 years ago there was a significant change in climatic patterns towards longer cycles with more marked shifts between warmer and colder conditions. Eight major glacial/interglacial cycles occurred, each lasting around 100,000 years. Ice built up gradually over 80–90,000 years reaching a peak before melting rapidly with the onset of interglacial periods which lasted for around 10,000 years. For most of the past million years then, glacial conditions have been the norm and the period of time during which the climate has been as warm or warmer than today has been very limited (Fig. 2.1). The last interglacial period occurred around 130–120,000 years ago. The present one has lasted for about 10,000 years. The last interglacial seems to have been a little warmer than the present one. On the basis of these 100,000-year cycles we are now moving towards the end of an interglacial. Left unhindered the Earth would probably start to move back towards glacial conditions some thousands of years in the future though human influences now look set to more than counterbalance such a trend.

How can these drastic climatic shifts be explained? For the last million years influences on global climate such as continental drift and mountain building have been negligible. The regularity of this cycle suggests that a recurrent forcing mechanism was at work. The glacial/interglacial sequence fits in well with the known pattern of variation of receipt of solar radiation caused by changes in the Earth's orbit.

The possibility that long-term variations in the Earth's orbit might have been responsible for the glacial/interglacial sequence was first suggested in the 1860s by the Scottish natural historian James Croll. The idea was refined by Milutin Milankovitch, a Serbian mathematician, in the 1920s and 1930s. He identified periodicities of around 22,000 years relating to the wobble of the Earth's axis (the precession of the equinoxes), 41,000 years related to variations in the tilt of the Earth's axis, and about 100,000 years as the planet's orbit changed from being circular to slightly elliptical and back again (Mintzer, 1992) (Fig. 2.2).

(a) the last million years

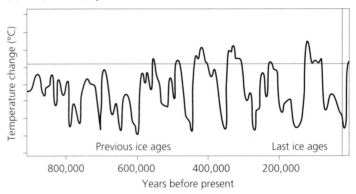

(b) the last ten thousand years

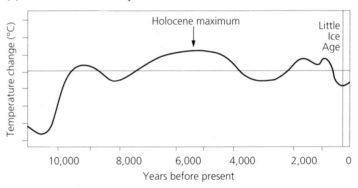

(c) the last thousand years

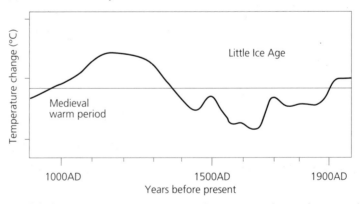

Fig. 2.1 Global temperature variations on three time-scales. (After Houghton et al., 1990)

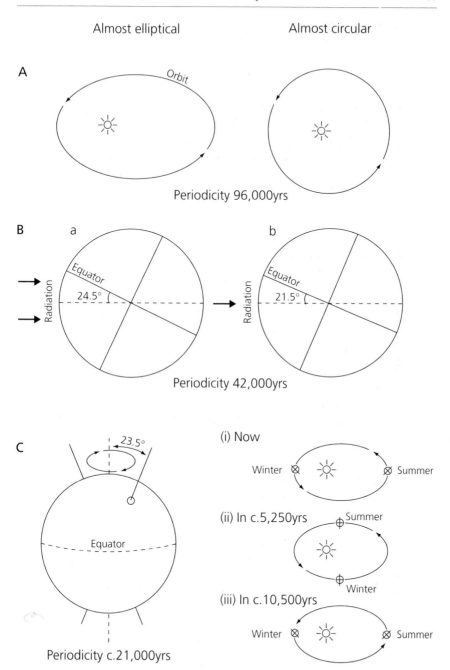

Almost elliptical   Almost circular

A

Orbit

Periodicity 96,000yrs

B   a   b

Equator   Equator
Radiation   24.5°   Radiation   21.5°

Periodicity 42,000yrs

C   23.5°   (i) Now

                          Winter ⊗   ⊗ Summer

             Equator   (ii) In c.5,250yrs   Summer

                          Winter

                          (iii) In c.10,500yrs

                          Winter ⊗   ⊗ Summer

Periodicity c.21,000yrs

Fig. 2.2 Variations in the Earth's orbit and axis affecting receipt of solar radiation. A. Eccentricity of the orbit. B. Tilt of axis. C. Precession of the equinoxes
due to wobble of the axis. (After Lowe and Walker, 1984)

Milankovitch's theories did not fit the glacial/interglacial chronology accepted in his day but more recent research has revised this chronology considerably and shown that it does fit the Milankovitch cycles quite closely (Emiliani, 1993). The 100,000-year cycle involved changes in the distance between the Earth and the Sun, thus affecting the amount of solar radiation received. The two shorter Milankovitch cycles involve no changes in the total amount of solar radiation reaching Earth, merely its distribution between the northern and southern hemispheres. The longest cycle could be considered as helping to drive the glacial/interglacial cycle while the shorter ones help to account for some of the briefer fluctuations between cold and warm conditions within major glacial phases (Fig. 2.3).

It seems that orbital forcing is a major factor in the timing of glaciations and deglaciations but exactly how it affected climate leading to ice sheet growth and decay is less clear. It has been suggested that the growth of ice sheets was encouraged when summer radiation receipt in the northern hemisphere was reduced due to orbital factors but when ocean temperatures in high latitudes were still warm, providing a source of abundant moisture which was unloaded on the cool continents as snow. The oceans eventually cooled but, by this time, the build-up of ice sheets had reached a point where it became a self-perpetuating mechanism. Deglaciation may have occurred when radiation receipt was high in summer and low in winter. Ice sheets began to melt and the North Atlantic became flooded with a surface layer of fresh water from calving icebergs. This froze in winter to form sea ice, preventing evaporation and

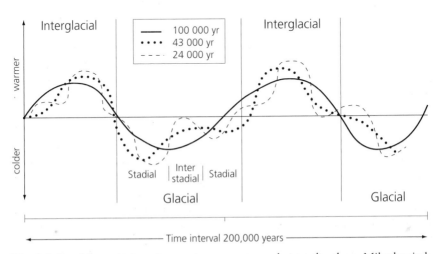

Fig. 2.3 Possible variations in mean temperatures due to the three Milankovitch cycles. (After Lowe and Walker, 1984)

cutting off snowfall to the ice sheets. As the ice volume was reduced, sea level rose encouraging more iceberg calving, yet another self-perpetuating process.

However, many scientists believe that the changes in the radiation balance involved in the Milankovitch cycles are not sufficiently great on their own to have caused the huge climatic shifts between glacial and interglacial conditions. The Milankovitch cycles may have been the background influence encouraging the growth or shrinkage of ice sheets but the link between ice ages and orbital variations is not quite so simple. The 100,000-year cycle does not involve the greatest changes in the distribution of solar radiation. These occur with the shorter periodicities of around 22,000 and 41,000 years (Liu, 1992). Changes in radiation receipt over the 100,000-year periods are small compared with the scale of climatic changes so that other factors operating within the Earth's environmental systems must have amplified the cycle's effects. Internal feedback mechanisms including variations in the atmospheric content of carbon dioxide and methane and fluctuations in the ocean circulation are now seen as important forcing factors (Peel, 1993). At the height of the last glaciation, carbon dioxide levels were only 70 per cent of those occurring in postglacial times before man began to modify the composition of the atmosphere significantly (Fig. 2.4). Variations on this scale could have played a significant part in the shift between glacial and interglacial conditions.

During the glacial/interglacial cycle, global mean surface temperatures may have fluctuated by around 5–7°C but in mid- and high latitudes of the northern hemisphere they may have varied by as much as 10–15°C. The onsets of glaciations have generally been gradual, but their terminations were rapid. Until recently it was suggested that average temperatures may have changed as quickly as 1°C per century and, for brief periods, even faster. Now, according to some suggestions, even more rapid shifts of up to 10°C within a couple of decades may have occurred. These temperature changes have been linked, by the analysis of bubbles of air trapped within ice cores from Greenland and Antarctica, to substantial fluctuations in levels of atmospheric carbon dioxide and methane.

Within the last 20 years, startling evidence has come to light that has caused a revolution in our thinking on the speed of climatic changes. It appears that glacial periods were not continuously cold but were interrupted by warm phases. Analysis of oxygen isotope data from the recent Greenland Ice Core Project (GRIP) has shown that around 24 brief phases of abrupt warming interrupted the last glaciation which extended from 115,000 to 13,000 years ago. During some of these warm phases temperatures may have altered as fast as at the end of the main glaciations. This has now been confirmed from high-resolution records of variation in planktonic foraminifera from North Atlantic sediment cores (Zahn, 1994).

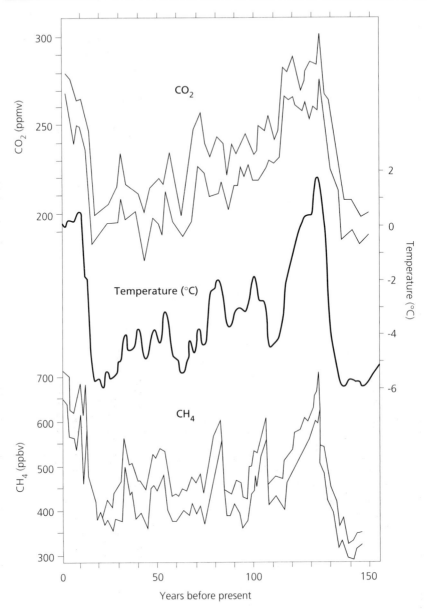

Fig. 2.4 Variations in the concentrations of carbon dioxide and methane against temperature changes over the last 150,000 years from the Vostok Antarctic ice core. (After Lowe and Walker 1984)

The abundance of foraminifera varies with ocean surface temperatures providing a link between the North Atlantic ocean circulation and the ice core evidence.

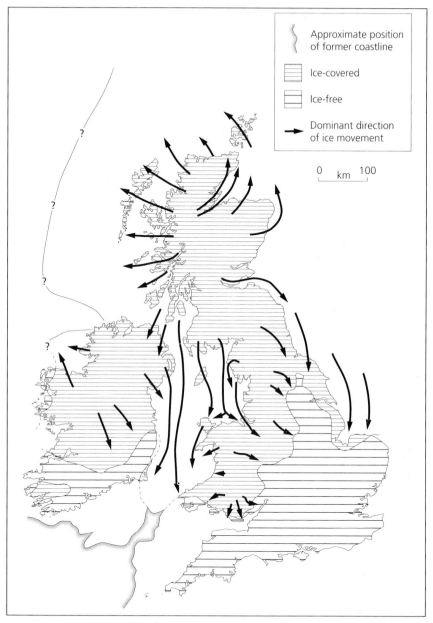

Fig. 2.5 Maximum extent of the last ice sheet in the British Isles. (After Lowe and Walker, 1984)

The last glaciation reached its maximum extent between 22,000 and 16,000 years ago. Ice sheets up to 4 km thick covered large areas including almost all of Scotland, most of Ireland and Wales and much of

Northern England (Fig. 2.5). Mean global temperatures may have been up to 8°C lower than now though in some areas they may have been as much as 20°C below those of today. Sea surface temperatures in the North Atlantic may have been around 12–13°C lower than they are now though in tropical waters like the Caribbean they were only 2–3°C less. In Britain mean July temperatures were around 10°C but in winter as low as –20 to –30°C (Fig. 2.6). Mean annual temperatures in Britain were probably between –8 and –12°C. In the tropics vegetation zones were 1–2,000 metres lower than at present. Most low and mid-latitude areas were significantly drier than now and fossil sand dunes dating from this time can be found underlying modern rain forests.

After about 16,000 years ago the ice sheets began to retreat and thin. Deglaciation took approximately 8,000 years to complete. At the end of the last ice age there were two massive flows of glacial meltwater into the oceans. The first, about 14,500 years ago, was caused by the collapse of the Eurasian ice sheet over northern Europe and Russia. The speed at which sea level rose at this time suggests that, rather than being due to the surface melting of the ice sheet, it was caused by changes in ice sheet dynamics, with an acceleration of ice flow towards the margins and massive calving of icebergs into the sea. The second huge discharge of meltwater occurred about 12,800 years ago due to the break-up of the Laurentide ice sheet over North America. It is not clear why the

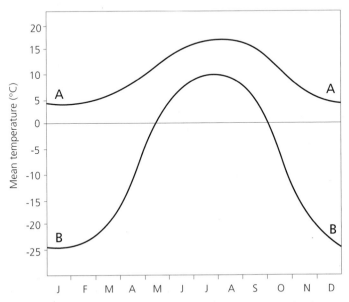

Fig. 2.6 Present mean monthly temperatures for central England (A) compared with probable trends at the height of the last glaciation (B). (After Lowe and Walker, 1984)

Laurentide ice sheet broke up so much later than the European one, whether due to different climatic conditions or to the American ice sheet having a higher tolerance to warming.

The decline of the ice sheets may have been due as much to aridity as to rising temperatures, for the main phase of ice decay in Europe occurred before a marked rise in temperatures occurred. The ice sheets created anticyclonic conditions which deflected away moisture-bearing winds. They may thus have expanded to a point at which they cut off their own supply of snow. At this time the Polar Front may have been well to the south of Britain, leaving dry, intensely cold conditions with little cloud cover and more solar radiation which would have encouraged the ice sheets to decay. Recent evidence suggests that, in Greenland, the change from glacial to interglacial conditions was abrupt, within five years or less (Fairbanks, 1993). The possible causes of such dramatic climatic shifts are considered below.

By around 14,000 years ago deglaciation was half complete with most of the European ice gone, but the southern margin of the North American ice sheet had retreated very little; the ice sheet had thinned rather than wasted back. Around this time temperatures in Britain rose abruptly with summer temperatures reaching levels similar or even higher than today. The evidence of fossil beetle assemblages shows that maximum warmth was reached about 13,000 years ago following rapid warming with mean temperatures rising possibly as much as 1°C per decade. Average July temperatures over much of England may have been in the region of 18°C. Interglacial conditions had arrived although, as it turned out, not for long.

THE YOUNGER DRYAS EVENT

The climatic amelioration was short-lived. Between about 12,800 and 11,500 years ago a short but severe cold spell occurred (Alley et al., 1993). It is widely known as the 'Younger Dryas event' after the arctic/alpine flower Dryas octopetala which staged a rapid comeback throughout Britain when tundra conditions returned. Temperatures may have dropped by 7°C to near-glacial conditions in under 50 years. Before the Younger Dryas event Scotland may have been completely deglaciated but, with the return of colder conditions, a major ice cap up to 600 metres thick built up over the central Scottish Highlands while small glaciers occupied valley heads in the Southern Uplands, the Lake District, Snowdonia and even the Brecon Beacons (Fig. 2.7). Mean July temperatures in southern England may have been as low as 10°C, in southern Scotland 8°C and mean January temperatures –17 to –20°C for much of Britain. The polar front, which had been pushed as far north as Iceland during the previous inter-stadial, now shifted south again towards Iberia. The end of the Younger

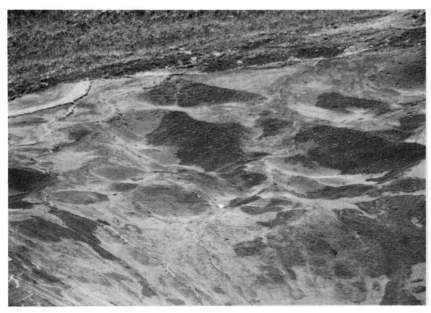

Fig. 2.7 Moraines from the Younger Dryas cold phase, Mickleden, English Lake District. (Photo: I. Whyte)

Dryas was equally abrupt; a northward shift of the Polar Front brought a doubling of snowfall within three years in parts of Greenland (Alley et al., 1993).

This cold phase was particularly marked in the North Atlantic area, affecting north-west Europe and eastern North America. There do not seem to have been major temperature changes in the Mediterranean but evidence from tropical areas suggests that the Younger Dryas was a widespread phenomenon. Cold conditions in the higher latitudes of the northern hemisphere were matched in East Africa by a contemporary phase of aridity, marked by abrupt falls in lake levels, suggesting a weakening of the Indian Monsoon circulation (Roberts et al., 1993). Climatic instability in the Andes is indicated by shifts between grassland and forest at high altitudes.

What caused this sudden but brief return to glacial conditions? A climatic change of this magnitude on such a short time-scale cannot be explained by variations in the Earth's orbit. The Younger Dryas episode occurred at a period of maximum radiation in the northern hemisphere. This cold phase is now being linked to changes in the circulation of the North Atlantic ocean. The pattern of interaction between climate, ice sheets and ocean circulation which is thought to have occurred during the Younger Dryas event also has implications for understanding the climatic variations which occurred during the last glaciation. It is thought that, at

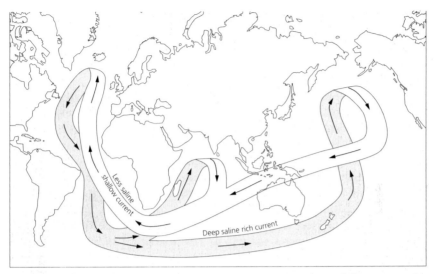

Fig. 2.8 The ocean 'conveyor belt' circulation driven by salinity differences. (After Lowe, 1993)

the start of the Younger Dryas event, the North American ice sheet experienced a massive collapse with large-scale calving of icebergs into the North Atlantic. This sudden release of meltwater from the Laurentide ice sheet caused a huge influx of low-density, cold, fresh water into the North Atlantic which blocked the northward spread of heat by cutting off the normal ocean circulation system. This provides a good example of the importance of links between the atmosphere and the oceans in determining climates.

Global ocean circulation is driven by differences in the salinity as well as the temperature of sea water. It is set up by salinity differences and maintained by heat exchange and density gradients. Quite small changes in salinity can produce large variations in ocean circulation and, as a result, climate. Surface water in tropical areas, which is warm and of low salinity, moves into the North Atlantic where it becomes cooler, denser and more saline. The Atlantic's high salinity is maintained by an excess of evaporation over rainfall. Water then sinks readily to the bottom of the ocean forming a current of 'deep water' which flows south around the tip of South Africa into the Indian Ocean and across the Southern Ocean to the Pacific. There it warms, rises, and is drawn westwards as a surface current across the Indian Ocean and back into the Atlantic, a circulation which is known as the 'Atlantic Conveyor Belt' (Fig. 2.8).

The sinking of surface water in the North Atlantic is caused by loss of heat and increased salinity due to evaporation. This is the mechanism which sets the conveyor belt in motion. Because it is driven by density differences and because saline water is more dense than fresh water, anything which decreases the salinity of North Atlantic surface water, such as increased precipitation, reduced evaporation or large-scale calving of icebergs, could stop the conveyor and cut the flow of ocean heat towards the pole, bringing cold conditions.

During the Younger Dryas episode the amount of fresh water released by the melting of the Laurentide ice sheet may have stopped the formation of deep water and reduced the contrast in density between surface and deeper ocean layers. The North Atlantic's vertical circulation slowed down, reducing the northward transfer of heat into the North Atlantic and encouraging the build-up of ice sheets. At this time there is evidence that the Gulf Stream flowed further south and that much of the North Atlantic was covered by sea ice. The timing of the onset and end of the Younger Dryas varied in different areas but, in Europe, glacier retreat began before temperatures started to rise markedly suggesting that, as during the previous glaciation, retreat was due to lack of precipitation rather than higher temperatures.

The Younger Dryas episode is important for a number of reasons. First, it illustrates the interrelationships which exist between the different elements of the climate system (see Chapter 3). Second, it demonstrates that major natural fluctuations in climate have occurred relatively recently within the present interglacial period. Third, it shows how sensitive climate can be to other environmental influences, emphasising that climatic change can occur quickly, within decades rather than centuries or millennia, as climate flips rapidly from one mode to another. Although warming rather than cooling is forecast for the future the warning still holds good; there may be thresholds within the climate system which, when crossed, can rapidly push climate into a very different mode, one less favourable to human existence.

Also during glacial periods the Atlantic Conveyor Belt may have been 'switched off' due to changes in salinity gradients. On a longer time-scale it is now being suggested that the ocean conveyor belt has undergone regular oscillations in the past on time-scales of around 1,000 years due to the increase and decrease of salinity in Atlantic surface waters caused in part by the temperature-controlled melting of ice sheets. These cycles punctuated the last glacial period. They seem to have been grouped into longer cycles of 10–15,000 years in which temperatures gradually fell with a series of colder interstadials, possibly reflecting the progressive strengthening of the polar cell. The cycles culminated in long, cold stadials terminated by brief, massive discharges of icebergs into the North Atlantic after which sea and air temperatures rose rapidly (Bond, 1993, Lehman, 1993 and Oerlemans, 1993a). The extent of the icebergs – up to 3,000 km across the Atlantic – has been plotted from layers of sediments rich in glacial deposits on the Ocean floor (Bond et al., 1992). In the North Atlantic the Polar Front, marking the boundary between warm and cold surface waters and the airflows associated with them, has operated like a door which shuts and opens with climatic changes to glacial and inter-glacial conditions. During the Holocene, the Polar Front has generally been north of Iceland, allowing warm subtropical waters to flow far to the north, driving the Atlantic Conveyor Belt's thermohaline circulation. During glacial phases the door was shut, with the Polar Front crossing the Atlantic directly towards Iberia, weakening the thermohaline circulation and preventing warm water from reaching into the North Atlantic. This situation was interrupted periodically during the last glaciation when the discharge of icebergs and meltwater destroyed the glacial circulation pattern and forced the door open for a while (Zahn, 1994).

It is not yet clear to what extent these longer cycles were due to exter-nal factors such as orbital variations or internal ones linked to ice sheet dynamics. It has been suggested that, in past glacial periods, ice sheets may have started out frozen to their beds, causing them to thicken rather than spread. A build-up of geothermal heat might eventually have brought the basal ice to melting point leading to a sudden collapse and outward surge of the Laurentide ice sheet every 10,000 years or so, much of it through the Hudson Strait into the North Atlantic. The ice sheet then refroze to its bed and the cycle started again. The release of a mass of water, perhaps equal to half the volume of the Greenland ice cap, into the North Atlantic within a relatively short period would have affected the ocean conveyor belt causing a short-lived but severe cold spell. The collapse of the ice sheet would then have led to the restoration of wind patterns in the North Atlantic similar to the present day which would eventually have set the conveyor belt in motion again bringing warmer conditions. A more recent interpretation is that the Atlantic Conveyor belt did not shut on and off but operated in three stable modes: one warm, deep mode similar to the present one; one much more cold and shallow, and one in between (Boyle and Weaver, 1994, Broecker, 1994 and Weaver and Hughes, 1994).

THE HOLOCENE OPTIMUM

The postglacial rise of temperature fluctuations in global mean surface temperature does not seem to have been outside a range of about 1°C on either side of the present average of 15°C, though greater variations did occur on a regional and local level. Because climate has been remarkably stable during the last 10,000 years it is easy to assume that major natural climatic changes are a thing of the past, something associated with the onset and break-up of glaciations but not relevant to modern conditions. In fact, the 10,000 years of relative stability in global climate which separates us from the Younger Dryas event is being seen more and more as an oddity rather than the norm. It is surely no coincidence that the rise of human civilization, based on the development of agriculture, has occurred during the longest period of stable climate of recent geological times.

Analyses of ice dating from the previous (Eemian) interglacial in the GRIP core from Greenland has produced disturbing evidence that, at this time, climate fluctuated between three distinct modes; one similar to that of the Holocene, another warmer and another cooler (White, 1993). It took less than a decade to shift from one to another and the different states were stable sometimes for thousands of years, sometimes only for decades. When analysis of the nearby GISP2 (Greenland Ice Sheet Project) core failed to provide evidence of the same changes during the last inter-glacial it was suggested that they had been due to the deformation of the layers in the GRIP core by ice movement. However, evidence from lake sediments in the French Massif Central has now confirmed that climatic changes did occur in the Eemian and affected continental Europe as well as Greenland (Thouveny et al., 1994). The data for the previous inter-glacial show that marked shifts of temperature accompanied by changes in greenhouse gas concentrations can also happen within interglacials. The worry is that greenhouse gas-induced global warming could take climate across one of these thresholds and set it into a quite different mode. Adapting to climatic change seems feasible enough when it occurs gradu-ally over several decades but is potentially much more traumatic if it happens within only a few years. At no time during the Holocene have Eemian-type climatic changes occurred but this is not to say that they could not happen in the future.

Nevertheless, the climatic changes which have occurred within the Holocene period, modest though they have been by comparision with the shift from glacial to interglacial conditions, have had profound environ-mental effects in some areas. In northern Mali, where rainfall is now only about 5 mm a year, remains of crocodiles and hippos have been found. Cave paintings from the central Sahara show antelope, giraffe, elephant and rhinoceros, among other animals. There was sufficient moisture earlier in postglacial times to permit rivers to flow from mountains like the Tibesti in the heart of the modern Sahara desert. The Sahara, which had been more extensive than today down to around 13,000 years ago, did not exist between about 9,000 and 5,000 years ago. A stronger monsoon circulation and a northward shift of tropical convectional rain brought much more moisture to the Sahara than it does now as well as to Arabia and north-west India, though areas like the Mediterranean, beyond the reach of the monsoons, may have been drier. As a result, much of the Sahara was savannah grassland rather than desert; Lake Chad at this time was as large as the Caspian Sea today. Lake levels in Africa continued high until 4,500–4,000 years ago when an arid phase set in. Since then there has been a steady decrease in lake size (Fig. 2.9).

Fossil beetle assemblages show a sudden rise in summer temperatures of around 7°C at the start of the Holocene in northern Europe. In north-west

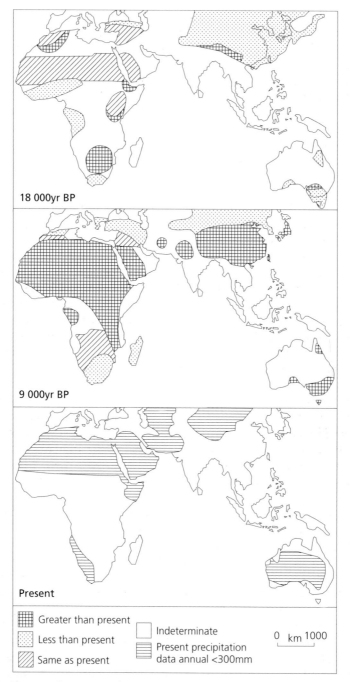

18 000yr BP

9 000yr BP

Present

Greater than present
Less than present
Same as present
Indeterminate
Present precipitation
data annual <300mm

0 km 1000

Fig. 2.9 Climatic changes in Africa, south Asia and Australasia since the last glacial maximum showing drier conditions in many areas at the height of the last glaciation and wetter conditions during the Holocene optimum. (After Roberts, 1994)

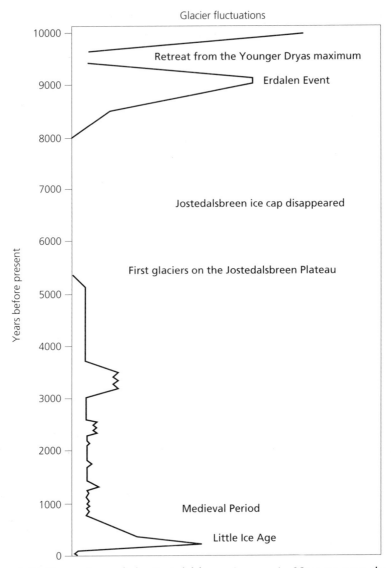

Glacier fluctuations

Fig. 2.10 Fluctuations of the Jostedalsbreen ice cap in Norway over the last 10,000 years. (After Mintzer, 1992)

Europe and many other parts of the world, modern summer temperatures were reached within the first 1,000 years of the Holocene. The return of warmth was slower in North America where the Laurentide ice sheet was still extensive until about 8,000 years ago when the sea invaded Hudson Bay and caused a widespread collapse of the ice.

The peak of postglacial warmth, the mid-Holocene climatic optimum, was reached around 7,000 years ago when mean temperatures, over the

North Atlantic and Europe at least, may have been around 2°C warmer in summer than today. This was enough to melt completely the Jostedalsbreen ice cap in Norway (Fig. 2.10). Tree limits were higher than they are now in Britain and the Alps; up to 200 m above those of recent times in Scotland, suggesting that mean temperatures were around 1.5°C higher. The northern limit of coniferous forest was 250 km beyond the modern limit in Canada and Scandinavia (Pastor, 1993).

It is not clear, however, to what extent this warmer phase was a truly global phenomenon though higher altitudinal limits to vegetation belts in New Guinea and South America suggest that warmer conditions affected parts of the tropics too. The postglacial climatic optimum has been detected in the Vostok ice core from the Antarctic but not as clearly as in the northern hemisphere. While much of northern and western Europe may have been 2°C warmer in summer and parts of Arctic Canada as much as 4°C warmer, the Mediterranean was cooler than today. The mid-Holocene climate in the northern hemisphere was characterised by enhanced seasonality and continentality due to the 22,000-year cycle of the precession of the equinoxes. At that time the Earth was closest to the sun in July rather than January as at present. As a result, the northern hemisphere received 8 per cent more radiation in summer than it does today. The more intense heating of continental interiors in summer created stronger monsoon circulation over Asia and Africa which brought rain farther north.

From about 5,000 years ago, cooler, less stable conditions began to affect many mid-latitude areas while parts of the sub-tropics became more arid. In the northern hemisphere temperatures fell by about 1°C between 5,000 and 3,000 years ago, causing the tree line in mountain areas to drop markedly. The northern limit of coniferous forest in Canada was pushed back between 200 and 400 km from its mid-Holocene maximum. The fall of summer temperatures in the Arctic may have been as much as 3–4°C. Cooling in the Arctic pushed depression tracks southwards across Britain so that conditions became wetter as well as colder (Harding ed., 1982). This encouraged the growth of peat in Britain's western and upland areas. At Tregaron Bog in Wales a metre of peat accumulated between around 800 and 400BC, as much as during the previous 2,000 years (Turner, 1981). As cooler conditions returned, some small ice caps which had melted during the postglacial optimum, such as the Meighen ice cap in arctic Canada, began to reform.

THE MEDIEVAL OPTIMUM AND THE LITTLE ICE AGE

Although global mean temperatures have only fluctuated slightly within historic times variations have sometimes been more marked at regional scales and have been sufficient to influence past societies. Conditions in the first centuries AD seem to have been more favourable than in later prehistoric times, for western Europe at least, but there was a renewed advance of glaciers in the Alps, Norway and Greenland in the seventh and eighth centuries. Ice blocked some passes across the Alps which had been used in Roman times. The medieval period, particularly the twelfth and thirteenth centuries in Europe, saw a temporary return of warmer conditions comparable with the mid-Holocene optimum. Tree lines were higher in the Alps, by 100–200 m, than in the seventeenth century. Warmer conditions encouraged the extension of settlement and cultivation to over 1,200 feet in hill areas of Britain such as Dartmoor and the Southern Uplands where traces of settlements and their field systems can still be seen. A northward shift of depression tracks and pack ice limits encouraged voyages into the Atlantic. First came Irish monks who settled in the Faeroes and Iceland in the eighth century. They were followed by Norsemen who colonised Iceland then went on to discover Greenland and the east coast of America as well as penetrating far into the arctic up the west coast of Greenland (McGovern, 1981).

Until recently it was believed that the medieval optimum was mainly confined to Europe. Some areas seem to have missed it or at least not experienced such a marked warm phase; China and Japan for instance. Elsewhere relatively warm conditions sometimes continued through from Roman times to the twelfth century. However, tree ring evidence from California, oxygen isotope data from cave deposits in Australia and patterns of glacier fluctuation in many other regions have shown that this warming phase affected many regions in both hemispheres, with mean temperatures about 1°C above those for the twentieth century. Many lakes, rivers and swamps in California are bordered by drowned tree stumps indicating former lower water levels. Dating of the most recent rings in these trees has indicated two periods of sustained, intense drought, around AD892–1112 and 1209–1350 (Stine, 1994). These are thought to have resulted from a contraction of the circumpolar vortex and possibly also a change in its pattern of ridges and troughs (see below) due to warmer global temperatures.

The medieval optimum came to an end shortly after about AD1100 in Greenland, but some 200 years later in western Europe (Fig. 2.11). Pack ice limits in the North Atlantic advanced south, cutting the sailing route from Iceland to the Norse settlements in Greenland. Beset by colder conditions which threatened their livestock and brought south Eskimos who attacked their settlements, the Norse colony dwindled. Some of their dead were buried in ground that is now permanently frozen. Eventually, perhaps in the fifteenth century, the colony died out altogether (McGovern, 1981). In Iceland the population faced a tougher existence as colder conditions threatened food supplies (Bergthorsson, 1985). During the medieval optimum, cereal cultivation had been possible in Iceland but now cooler, wetter summers prevented grain from ripening.

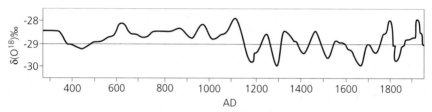

Fig. 2.11 Temperature changes shown by oxygen isotope variations from north-west Greenland since AD300. The relatively high temperatures of the medieval optimum and the sharp fall at around 1100 at the start of the Little Ice Age are especially clear. (After Lamb, 1982)

After a run of favourable harvests, colder, wetter conditions brought severe famine to much of Europe in 1315–17 (Kershaw, 1973). Glaciers began to advance in the Alps in the late thirteenth and early fourteenth centuries and at almost the same time in the Himalayas and New Zealand (Grove, 1988). The later thirteenth and fourteenth centuries brought stormier conditions around the North Sea leading to more frequent severe flooding of coastal lowlands in eastern England and the Low Countries, a pattern which continued throughout the fifteenth century. The medieval warm period was succeeded by a colder phase which has become known as the Little Ice Age (Grove, 1988). It was first identified in Europe from evidence of glacier advances in the Alps and Scandinavia between the sixteenth and nineteenth centuries (Le Roy Ladurie, 1971). The name 'Little Ice Age' is misleading though; it suggests that there was a marked cold interval which was synchronous over a wide area. In fact climate variations over the last 500 years or so were much more complex, more varied regionally, and harder to define. Extreme conditions were more frequent with unusually hot or wet summers as well as cold winters (Grove, 1988). The Little Ice Age was not continuously cold. After its start in the thirteenth and fourteenth centuries there was an interval of more favourable conditions in Europe before a return of more severe weather between the mid-sixteenth and mid-nineteenth centuries (Grove, 1988). However, this broad outline chronology varied in timing from one part of the world to another and the cooler phases within it did not necessarily occur simultaneously in different regions. Some synchronous cold spells are recorded at various sites through the northern hemisphere but they tend to be short – generally 30 years or less – such as the 1590s–1610s, 1690s–1710s, 1800–1810s, 1880s–1900s. The Little Ice Age does seem to have been a worldwide event though for it has been picked up in data from North and South America, China and Japan as well as Europe.

There is no agreement on when the Little Ice Age started, when it ended, or even what its basic climatic features were. Its beginning in western Europe has been placed as early as 1250 and as late as 1550. Even for central Europe the climatic record shows no clear sign of a prolonged cold phase. Climate seems to have been more variable than in recent times though there was a marked shift to colder and wetter summers in the last third of the sixteenth century (Pfister, 1985). In northern Italy the Little Ice Age was heralded by an increase in the number of severe winters in the fifteenth century compared with the fourteenth. At Camp Century in Greenland and Devon Island there was a long cool period from around 1560–1850 preceded by an earlier downturn of temperatures in the later thirteenth century. In parts of the southern hemisphere cooler conditions set in as early as the sixteenth century. The impact of the Little Ice Age was especially severe in marginal upland areas. Cooler, wetter conditions drove cultivation limits in western Europe steadily downhill between the fourteenth and late seventeenth centuries, encouraging the abandonment of high-lying settlements and their fields in parts of Britain and central Europe (Parry, 1976 and 1977).

The nadir of the Little Ice Age in many parts of Europe was the 1690s when mean temperatures may have been 1°C lower than in the first half of the twentieth century. Cool wet summers brought crop failure and famine to Scotland and Scandinavia. At the end of the 1690s summer sea ice limits lay south of Iceland. There are records of polar bears being washed ashore in the Faeroes and strange men in single-seat skin boats – apparently Eskimos – being sighted off Shetland and Orkney (Lamb, 1977). The seas north of Scotland may have been 5°C colder at this time than they are today. Glaciers advanced in the Alps and Scandinavia while tree lines may have dropped 100–200 m in middle latitudes and 300 m in equatorial areas (Pfister, 1981 and Grove, 1985). In Scotland large areas of permanent snow became established on the highest mountains and there have even been suggestions that small glaciers began to form in the Scottish Highlands (Grove, 1988) (Fig. 2.12). In the late seventeenth and early eighteenth centuries Norwegian glaciers were threatening farms and causing their inhabitants to petition for reductions in taxes.

Following the 1690s there was a sharp rise in mean annual temperatures in the early years of the eighteenth century in England and then a much more gradual rise, but relatively cold conditions persisted until the later nineteenth century. In western Europe the nineteenth century seems to have been an especially cold period within the Little Ice Age. In Iceland the Little Ice Age is better defined as lying roughly between 1750 and 1900; maximum glacier positions were reached in the mid-eighteenth century with glaciers remaining advanced for about 150 years. The low point of the Little Ice Age in central Canada was the end of the 1760s. The ice core from the Quelccaya ice cap in Peru shows that, in terms of dust content and oxygen isotope values, the Little Ice Age began and ended abruptly with a sharp change occurring within three years, around 1490 and 1880 (Bradley and Jones, 1992) (Fig. 2.13).

Fig. 2.12 Snow patches on the north face of Ben Nevis. During the Little Ice Age there may have been tiny glaciers here. (Photo: I. Whyte)

Quelccaya, Peru

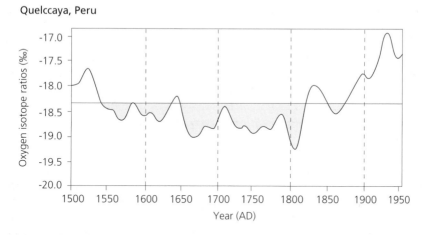

Fig. 2.13 Isotope values (and temperatures) for the Quelccaya ice cap, Peru, showing the abrupt start of the Little Ice Age, around AD1490, and its ending in about 1880. (After Bradley and Jones, 1992)

Despite its variability and uncertain chronology, the Little Ice Age was a truly global phenomenon and may have been the most widespread cold phase since the end of the Younger Dryas. It has, indeed, been considered by some as an abortive glaciation which did not last quite long enough to trigger off a major ice advance (Andrews et al., 1976). In the view of some it could have represented the initiation of post-Holocene cooling towards the next glaciation whether interrupted by human intervention or natural variability (Kullman, 1994). Its causes, however, remain elusive. Variations in atmospheric and ocean circulation in the North Atlantic area have been suggested (Grove, 1988) as well as changes in solar radiation. One recent suggestion is that changes in the circulation of the Pacific Ocean were responsible but studies of glacier fluctuations in northern Scandinavia have revealed other similar cold phases within the last 10,000 years which may be linked to small variations in the output of solar radiation. Atmospheric cooling due to the short-term effects of major volcanic eruptions may also have played a part (see below).

In the past historians have tended to dismiss the influence of climatic change on human activity. The French historian Le Roy Ladurie (1971), having produced a meticulous chronology of the Little Ice Age, concluded by suggesting that a drop in mean temperature of 1°C was unlikely to have had any significant effect on agriculture or other economic activities. While bearing in mind the dangers involved in being too superficial in attributing historical events to climatic causes, the view that climatic changes have had a more widespread influence on man in recent centuries is now more widely accepted. Given the problems experienced by the inhabitants of many areas as a result of the post-medieval climatic downturn it is worth reflecting that, in terms of average global temperatures, only 1°C separates us from the nadir of the Little Ice Age, though the scale of cooling at a regional level may have been 2–3°C in some areas.

TWENTIETH-CENTURY WARMING

If there is doubt about the timing of the start of the Little Ice Age, the time of its end is quite clear. Sustained warming only began at the end of the nineteenth century, accompanied by a worldwide retreat of glaciers (Fig. 2.14). For the twentieth century, increasingly large databases have been assembled for weather stations around the world by the Climatic Research Unit at the University of East Anglia. Data from almost 2,000 land-based stations, plus ship-based observations of air and sea surface temperatures have been combined. They show a warming trend which has been marked but is not uniform through time or between different regions (Jones, Wigley and Wright, 1986). Since the mid-nineteenth century, average global temperatures have risen by up to 0.6°C or, if 0.1°C is subtracted to allow for purely local warming influences such as the impact of urban climates, 0.5°C (Fig. 2.15). Trends in sea surface temperatures

Fig. 2.14 Nigardsbreen glacier, Norway. The Little Ice Age limits show up as a trim line on the hillside above the present glacier. (Photo: Dr. P. Vincent)

have been similar to those in the atmosphere, warming globally by around 0.4°C since the late nineteenth century.

With the post-1970s warming, northern snow cover has receded – in North America by up to 10 per cent – to a record minimum in 1990. The fall of temperatures due to the Mount Pinatubo eruption of 1991 has restored snow cover to its 1970s extent but this may only be a temporary

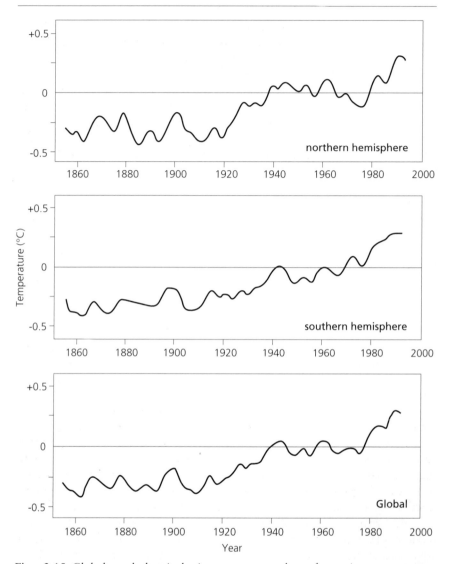

Fig. 2.15 Global and hemispheric mean annual surface air temperatures 1854–1992, shown in relation to the 1951–80 average, based on both land and marine data. (After Roberts, 1994)

reversal if the warming trend continues (Cohen, 1994). There have also been reports of a marked thinning of sea ice in parts of the Arctic but the high degree of interannual variablity of ice thickness and the limited number of observations makes it impossible so far to establish any trends with confidence (McLaren et al., 1992).

An increasing number of regional studies have confirmed the global warming trend from areas as far apart as the Antarctic Peninsula (King, 1994), the Middle East (Nasralla and Balling, 1993) and China, where a rise in mean annual temperatures of up to 0.4°C has been recorded within the last 40 years (Hulme et al., 1994). While the warming has been a global phenomenon there have nevertheless been contrasts between the northern and southern hemispheres. The rise of temperatures in the southern hemisphere has been fairly regular. In the northern hemisphere, however, much of the increase occurred in the late nineteenth and early twentieth centuries and, in recent years, with a period between the 1940s and the 1970s when average temperatures dropped slightly.

This cooling phase now seems to have been short-lived but it caused considerable concern at the time. In Iceland, a country whose economy is especially sensitive to slight climatic shifts, cooling during the 1960s brought sea ice back into Icelandic waters, reduced the hay crop, affecting livestock production, and damaged the herring fishery to an extent which had major repercussions on the Icelandic economy. The drop in temperature was especially marked in parts of the Arctic while even in England the growing season was cut by nine–ten days in the 1960s compared with earlier decades. As discussed in Chapter 3, this temporary cooling phase may reflect the impact of human influences such as the release of aerosol particles which may have partly counterbalanced the effects of greenhouse gas increases.

An interesting discovery is that recent global warming has occurred mainly at night. Since the 1950s, minimum daily temperatures over most of the land areas of the northern hemisphere have risen about three times as fast as maximum ones. Nights are on average 0.84°C warmer, days only 0.28°C, than a century ago. This applies to all seasons and all continents. It does not seem to be the result of urban heating effects because the trend is still evident even when data from large towns and cities are omitted from the calculations. It is possible that aerosols, due to industrial pollution, have reduced the impact of greenhouse gas forcing during the daytime. Cloud cover seems to have increased along with warmer nights. Greater low-altitude cloud cover may be trapping more heat close to the Earth's surface by night but reflecting more incoming radiation by day.

The twentieth-century temperature rise has been linked to widespread glacier retreat. The ice core from Devon Island in arctic Canada, for example, has melt features within the ice which suggest that there has been a negative mass balance over the last century compared with a positive one for the previous 300 years of the Little Ice Age (Bradley and Jones, 1992). In recent years the rate of retreat of some glaciers has slowed, perhaps as a delayed response to cooling in the 1960s and early 1970s, though with a tendency for the percentage of glaciers in the Alps which were retreating to rise markedly at the end of the 1980s.

Central England temperatures in the last 30 years have been 0.5°C above the mean for the entire time series extending back to 1659 and 1°C warmer than the coldest period at the end of the seventeenth century. The warmest year in the whole record to date was 1990. In terms of global averages the six warmest years of the twentieth century occurred in the 1980s, with 1987 and 1988 being the warmest on record. This pattern has continued into the 1990s with 1991 setting new records. Global average temperatures in the 1980s were higher than in any decade in the last 140 years. This warming trend has been

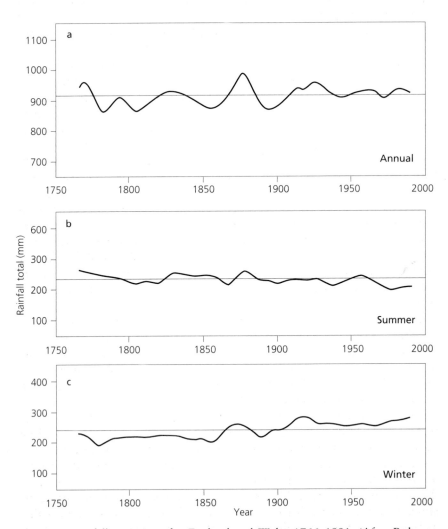

Fig. 2.16 Rainfall variations for England and Wales 1766–1991. (After Roberts, 1994)

punctuated by the effects of the eruptions of El Chichon in Mexico (1982) and Mount Pinatubo in the Philippines (1991). Both eruptions caused short-term cooling worldwide as a result of the dust and gases they released into the atmosphere (see below). The Pinatubo eruption brought the welcome return of good snow conditions to ski resorts in the Alps, the Pyrenees and the Cairngorms, after several years of limited snow cover.

The scale of the temperature rise that has occurred within the last 150 years has been within the range which, it has been calculated, should have occurred due to the increase in greenhouse gas concentrations over the same period, though admittedly towards the lower end of the range. On the other hand, global mean temperatures during the past few thousand years have gone up and down by as much as 1°C so that the recent increase is still well within the range of natural variability.

Computer models suggest that global precipitation should increase by 2–3 per cent for each 1°C rise in mean temperature. Due to problems of measurement it is not possible to identify any such rise in precipitation over the last century or so. Precipitation changes at global and even regional levels are harder to measure than temperature variations (Fig. 2.16). There are no data for the oceans or from satellites and there is a high degree of local variability which makes generalisation difficult. Global-scale analysis of land-based rainfall data shows no significant trend upwards or downwards and no indication of a change in the variability of precipitation. On the other hand, there have been marked changes in some regions; an increase in the mid-latitudes of the northern hemisphere and a decrease in subtropical areas. In Russia annual precipitation has increased fairly steadily in the last century. In the Sahel there has been a substantial drop since the 1960s – in many areas of 30 per cent or more – to levels much lower than during the first half of the twentieth century (Fig. 2.17).

There has been a general perception in many parts of the world that weather conditions have been more variable in recent years than in the past with a greater incidence of extreme conditions. It is not easy, however, to link these specific events to the more general warming trend or indeed to show conclusively that the temperature rise has itself resulted from the addition of greenhouse gases to the atmosphere by man. On the other hand, the scale of the recent warming trend, and some regional patterns of climatic change such as drier conditions in the American Midwest and Great Plains, are consistent with computer-modelled scenarios of greenhouse gas-forced climate change over the same period (Chapter 4). While many scientists are convinced that this warming trend, especially the acceleration in the 1980s, is the first manifestation of anthropogenic global warming the verdict is still 'not proven'. The consensus view is that at least another two decades of research will be necessary before climatologists can be certain.

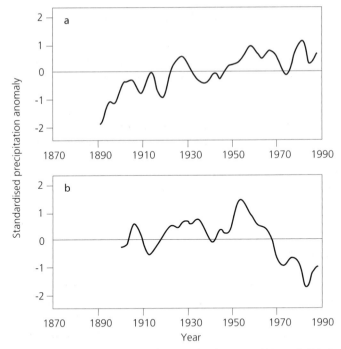

Fig. 2.17 Variations in precipitation for (A) the former USSR and (B) the African Sahel, expressed in terms of standard deviations away from the mean. (After Roberts, 1994)

VARIABILITY WITHIN THE PRESENT CLIMATE SYSTEM

It is assumed that the reader has some basic knowledge of the atmosphere, its circulation, and the resulting patterns of weather and climate. However, to understand some of the reasons for the year-to-year variability which occurs under present conditions it is worth focusing briefly on three different types of influence; volcanic eruptions, changes in the upper westerly circulation, and the El Niño/Southern Oscillation phenomenon.

One of the problems with trying to measure climatic change on a time-scale of decades or longer is that long-term trends, the 'signal', are hard to detect because they are obscured by the background 'noise' caused by such short-term variations which can be far larger in scale. Such factors will, of course, continue to influence climate even if global warming does occur, sometimes working with it, sometimes against it.

Volcanic eruptions

One of the most spectacular short-term effects is the impact of major volcanic eruptions (Chester, 1988). It has long been known that these can have an effect on climate. The eruption of Tambora in the Dutch East Indies in 1815 blasted at least 15 km³ of material into the atmosphere and produced a sudden global cooling which, in Europe, led to 1816 becoming known as 'the year without a summer'. Poor grain harvests throughout Europe led to large-scale famine, the 'last great subsistence crisis' of Europe as it has been termed (Post, 1977). There is also growing evidence that major volcanic eruptions in prehistoric times produced short-term changes in climate which had severe impacts on ecosystems and human societies (Blackford et al., 1992 and Burgess, 1989).

Recent studies have shown that major eruptions can lower global temperatures for two or three years afterwards. There have also been suggestions that volcanoes may have had longer-term impacts on temperature (Bradley and Jones, 1992). Study of this is complicated by the likelihood that many significant eruptions during the past few centuries in remote areas such as the Aleutians or Kamchatka may not have been recorded. This can happen even in modern times. The eruption of the volcano Itiboleng in Indonesia in 1983 went unreported until it was observed by the crew of a Space Shuttle (Grove, 1988). In addition, the impact of known eruptions on climate is not easy to estimate accurately because of the difficulty of assessing their strength and nature. It is now known that the climatic impact of an eruption depends not only on the volume of material ejected from the volcano and how high it is thrown into the atmosphere, but also on the location of the volcano in relation to atmospheric circulation, and the chemical composition of the ejected tephra and gases.

It used to be thought that major eruptions were associated with cooling because particles of volcanic dust were thrown high into the atmosphere blocking solar radiation. In 1970 Lamb produced a Dust Veil Index (DVI) which attempted to assess the global climatic impact of all major eruptions since AD1500 (Lamb, 1982). Starting with the well-recorded eruption of Krakatoa in 1883, he assigned it a DVI value of 1000 and calibrated all other eruptions relative to it. The Index was not completely objective, however, and was biased towards the effects of eruptions on the midlatitudes of the northern hemisphere.

Lamb also produced a cumulative index in which dust from individual eruptions was spread out over four-year periods to simulate its gradual fallout from the atmosphere. The index suggested that, relative to the twentieth century, the preceding four centuries, spanning the Little Ice Age, had been ones of comparatively high atmospheric volcanic dust content, prompting speculation as to whether the warming of the later nineteenth and early twentieth centuries was linked in part to an absence of major eruptions.

A more refined chronology, the Volcanic Explosivity Index (VEI), has been developed, based primarily on geological criteria with eruptions graded from 1 to 8 (the most severe) (Bradley and Jones, 1992). Volcanoes with a VEI of 4 or more were powerful enough to have ejected material into the stratosphere and to have influenced climate at more than a regional scale. In the last 500 years there have been about 110 eruptions of this magnitude.

More recently it has been realised that emissions of sulphur dioxide gas from eruptions also affect climate (Bluth et al., 1993). The sulphur dioxide is converted into tiny droplets of sulphuric acid which can block incoming radiation and reduce atmospheric temperatures. Satellites can now detect and track the progress of sulphur dioxide clouds from eruptions, allowing their impact on the atmosphere to be estimated. The eruption of El Chichon in Mexico in 1992 was the first for which gas emissions were monitored by satellite; around 10 million tons of sulphur dioxide were released.

The problem with assessing the climatic impact of volcanic eruptions is that the most violent explosions do not necessarily have the biggest effects on climate. In terms of gas emissions, the famous Krakatoa explosion of 1883 was not especially severe. In terms of output of sulphuric acid the worst eruption of the last 500 years was at Laki in Iceland in 1783, with only a moderate VEI of 4. This eruption may have produced more sulphuric acid than Tambora, thought to have been the most violent eruption of postglacial times, but because the Laki eruption was not as violent, far less sulphur dioxide reached the stratosphere and although Europe experienced some anomalous weather as a result of low-level atmospheric pollution there was not a marked global-scale cooling. Low-latitude eruptions like Tambora's tend to have a disproportionately great effect on climate because their dust and gas is carried around the world by atmospheric circulation while major eruptions in high latitudes are often less effective because the ejected material is not spread as widely.

In 1991 Mount Pinatubo in the Philippines ejected 20 million tons of sulphur dioxide into the stratosphere leading to measurable global cooling. Non-explosive volcanoes like Etna produce a steady release of sulphur dioxide into the atmosphere estimated at about nine million tons a year, while the sporadic output from major explosive eruptions averages out at only around four million tons a year. Past sulphur-rich eruptions can be detected from acid fallout preserved as layers in ice cores. Studies of cores from Greenland have shown that the quietest period of volcanic activity was from around AD1100–1250, coinciding with the medieval optimum and the most active periods 1250–1500 and 1550–1700, suggesting strong links with the Little Ice Age (Grove, 1988).

Even more disturbingly there is now evidence that a massive eruption at Toba in Sumatra about 63,500 years ago may have helped to plunge the Earth into the last glaciation. At this time the climate of the northern hemisphere had already started its transition towards glacial conditions. The eruption, estimated as having ejected five times as much material as Tambora, may have cooled summers in high northern latitudes by 10–15 degrees and led to snow accumulating over the plateaux of Quebec and Labrador, triggering off a sudden plunge into colder conditions (Rampino and Self, 1992).

The upper westerly circulation

The core of the upper westerly circulation, or circumpolar vortex, is a belt of high-speed winds in the upper troposphere, the jet stream, about 8–10 km above the Earth's surface. The jet stream is steered north and upwards by the barrier of the Rocky Mountains and then south and downwards over the American Great Plains. This wave generates a series of further waves downstream around the northern hemisphere. These wave patterns vary on different time-scales. The circulation can be very fast and stream-lined, with only three shallow waves contracted towards the pole, or slower and more sluggish with four or five waves of much greater ampli-tude forming a series of great meanders (Fig. 2.18). This pattern of airflow in particular brings markedly different sets of weather conditions to differ-ent areas depending on whether they are dominated by a ridge of warmer air pushing northwards or a trough of polar air pushing southwards. Sometimes a meander may be cut off, creating blocking anticyclones which can remain almost stationary and persist for months, diverting the westerly circulation to north and south, conditions which often bring very dry

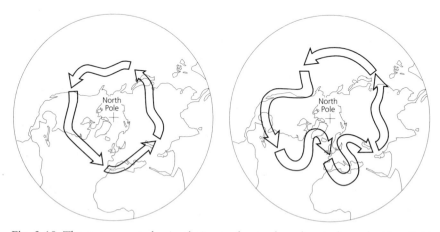

Fig. 2.18 The upper westerly circulation in the northern hemisphere showing (left) a faster, more streamlined pattern in three shallow waves and (right) a more sluggish one with five waves. (After Lamb, 1982)

summer weather to western Europe – as in 1976, 1984 and 1989 – or very cold conditions in winter.

Early climatic historians, not appreciating the characteristics of the upper westerly circulation, often failed to realise that the occurrence of markedly contrasting weather conditions in two different areas was due to the same basic cause, the configuration of the circumpolar vortex. Warm dry summer periods over most of Britain, due to the dominance of high pressure systems with blocking anticyclones, are often matched by wetter than average conditions in the northern and western islands of Scotland or in European Russia as depression tracks are steered north of Britain around the high pressure zone.

Variations in the upper westerly circulation in the northern hemisphere demonstrate the complexity of climatic change, the difficulty of reconstructing past variations, and the problems of forecasting the future. The strength of the vortex and the number, amplitude and location of the ridges and troughs within it, vary through time affecting the areas influenced by warm sub-tropical and cold polar airflows and the average positions of the low-level weather systems which they generate and steer. Closely related to them in the North Atlantic is the boundary between the warm water of the Gulf Stream and cold polar waters. At the climax of the last glaciation, this lay almost as far south as the Azores. As climate warmed, the boundary shifted progressively northwards until, at the time of the mid-Holocene optimum, it lay between Iceland and Greenland. During the cold 1690s, however, the boundary was pushed south of Iceland again almost into Scottish waters.

El Niño

El Niño the phenomenon known as 'the Christchild', as it generally occurs around Christmas, is a regional-scale manifestation of a large-scale ocean/atmospheric fluctuation affecting tropical areas of the Pacific. These larger-scale events are often termed El Niño/Southern Oscillation (ENSO) events (Philander, 1990). Under normal conditions in the tropical and sub-tropical Pacific there is a flow of air at high level from the area of low pressure over Indonesia eastwards across the Pacific and a low-level return airflow across the Pacific from east to west. Descending air over the eastern Pacific keeps rainfall low over the west coast of South America. When these strong, persistent trade winds blow, westward-flowing ocean currents are intensified, pushing warm water across the Pacific Ocean to build up on its western side. Periodically, however, the winds weaken, or even reverse into an eastward flow, causing a backwash of warm surface water into the seas off the coast of South America. An ENSO event sets in and, as the warm water spreads eastwards, the equatorial low pressure

centre shifts eastwards too, as do the areas of heavy precipitation which
are displaced towards the mid-Pacific and sometimes to the eastern Pacific
bringing torrential rain and flooding to the normally dry west coast of
Peru (Fig. 2.19). The 1982 El Niño event, the most severe on record,

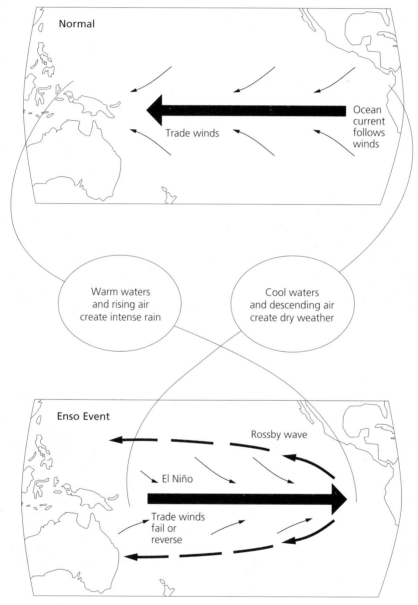

Fig. 2.19 How wind patterns and ocean currents in the tropical Pacific are affected
by ENSO events. (After Schneider, 1990)

brought major floods to Ecuador and Peru and widespread drought and forest fires to Australia and Indonesia (Allan, 1988 and Salafsky, 1994).

ENSO events have an impact on climates in other regions too (Glantz et al., 1991). The 1982–3 event was linked with persistent droughts in north-east Brazil, south and east Africa and in northern India. It is believed that ENSO events can cause jet streams to steer in an irregular pattern, producing weather anomalies in various parts of the world outside the tropical Pacific (Yarnal, 1985). A change in the jet stream in spring and summer 1988, which carried rainfall away from the USA to Canada causing drought, was thought to be due to a severe ENSO episode. ENSO events are linked to unusually wet weather in the south-west USA and the opposite phase of the cycle – La Niña – with extremely dry conditions in the central USA, as in 1988 (Keppenne and Ghil, 1992). The link between ENSO events and the variability of monsoon rains is now allowing more accurate prediction of rainfall patterns and crop yields in countries as far away as Zimbabwe (Cane et al., 1994 and Rosenzweig, 1994).

ENSO events are the most significant source of year-to-year variability in weather around the world. There are indications that ENSO events have not been a feature of climate throughout the postglacial period but may have begun around 4,500 years ago (Diaz and Markgraf, 1992). Since their occurrence can first be identified from historical sources in the early sixteenth century they have occurred on average about every four years. However, they vary greatly in intensity and the more severe ones only occur every nine years or so. In recent years the construction of a chronology of El Niño events from South American documentary sources as well as instrumental data has allowed the identification of periods when El Niño episodes were particularly frequent or strong (Bradley and Jones, 1992). A substantial increase in precipitation in north-west Peru between 1864 and 1891 may have been linked to the general global warming at the end of the Little Ice Age. Since the 1980s severe ENSO events seem to have become more frequent for reasons which are at present unknown, whether related to rising greenhouse gas concentrations or other influences. The potential impact of a change in the pattern and intensity of ENSO events is a major concern to climatologists around the Pacific (Colls, 1993).

The warming which has occurred in the last 150 years is unlikely, however, to have been due to changes in the pattern of eruptions, shifts in the westerly circulation, or the effects of ENSO events. Although it cannot yet be proved conclusively, the man-enhanced greenhouse effect seems the most likely cause of the twentieth century global temperature rise. The next chapter deals with the greenhouse effect, greenhouse gases, and how man has altered the composition of the atmosphere.

3

The Greenhouse Effect and Greenhouse Gases

Although it is assumed that readers have some background knowledge of the workings of the climate system, a few points need to be emphasised before looking at the greenhouse effect in detail. To understand how climate works, the atmosphere cannot be considered alone; the other environmental systems with which it is linked or coupled must also be taken into account (Fig. 3.1). Climate is the result, not only of complex sets of interactions within the atmosphere, but also between the atmosphere and the oceans, land surfaces and their vegetation, and snow and ice cover. Together these form the climate system. The Earth's climate is driven by energy from the Sun. The temperature of the Earth adjusts itself to maintain a balance between incoming and outgoing radiation but this balance can be altered by various **forcing factors**, the intervention of which can cause a readjustment of temperatures which can, in turn, affect the entire climate system. Changes in the amount of solar radiation received by the Earth, or variations in its seasonal distribution due to orbital changes, are two forcing factors. As discussed in Chapter 2, changes in the Earth's orbit are likely to have been a background cause of ice ages. There is currently considerable debate about the amount of variability of solar radiation on shorter time-scales and the effect that such changes may have on climate. Records of solar irradiance from satellite monitoring suggest that the amount of variation is only around 0.1 per cent, insufficient to have a significant effect on climate. However, such measurements have only been available for a few years and the possibility of longer-term variations which might have a greater impact on climate cannot be dismissed (Schonwiese et al., 1994).

The injection of dust and gas into the atmosphere from volcanic eruptions is another forcing factor as is change in the albedo (reflectivity) of the Earth's surface due to influences such as seasonal changes in snow and ice cover. Different kinds of surface absorb different amounts of radiation. Ice and snow cover reflect most of it but darker surfaces absorb more

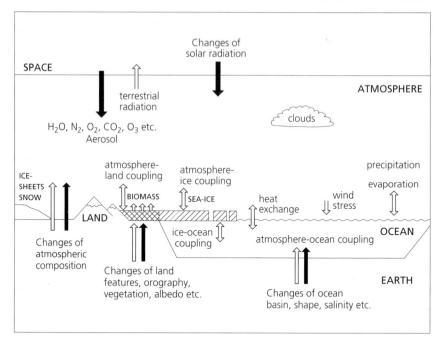

Fig. 3.1 Simplified diagram of the components of the coupled atmosphere-ocean-ice-land climatic system. Shaded arrows represent external processes, unshaded ones internal processes causing climatic change. (After Houghton et al., 1990)

forcing factors cont.d..

radiation. Any change in the albedo of the land surface due to the clearance of forests or desertification will affect the amount of solar radiation absorbed by the Earth's surface. The addition of greenhouse gases to the atmosphere as a result of human activity is another forcing factor.

The elements of the climate system have different response times to change. The atmosphere is the most variable element of the climate system, changing rapidly on time-scales of days and weeks. The ocean responds much more slowly to change; its surface layers vary on time-scales of months and years, but it may take centuries for changes to affect the ocean deeps. The oceans can absorb a lot of energy before experiencing a rise in temperature (Weaver, 1993). This thermal inertia has a major role in slowing the rate at which the atmosphere may warm in the future. Ice and snow cover has large seasonal variations; about 8 per cent of the Earth's surface is covered permanently but snow and sea ice cover causes this figure to double seasonally. In the northern hemisphere, only 4 per cent of the area is covered permanently but, in winter, this increases to 24 per cent. Snow cover has a very short response time to climatic changes but glaciers respond much more slowly while the Antarctic ice cap takes millennia to react to temperature changes. The position of the continents

response times

ice + snow

and their topography changes and influences climate on time-scales of millions of years but the biosphere can change much more rapidly as vegetation belts expand and contract and as natural vegetation is altered and cleared by man.

An important aspect of climatic change is the concept of feedback. As all elements within the climate system are interlinked, changes in any of them can have repercussions right through the system. Such knock-on effects may amplify the original disturbance or reduce it. Changes of this kind are known as feedback mechanisms. **Positive feedback occurs where change destabilises climate and pushes it into a different mode**, while responses which tend to damp down the original disturbance and stabilise the climate once more are termed **negative feedback**. An example of positive feedback is the ice-albedo mechanism where an initial expansion of snow and ice cover leads to more radiation being reflected. This cools the local or regional climate and encourages more precipitation to fall as snow. This then further increases the albedo and so on. Examples of negative feedback in the glacial/interglacial cycle have already been noted. At some point the build-up of ice reaches a **threshold** where cold conditions reduce precipitation, starving the ice sheets and leading to a thinning of the ice. Changes in the concentration of greenhouse gases in the atmosphere, especially carbon dioxide, may be associated with a range of positive feedbacks which could enhance global warming. This in turn may be at least partly compensated by some negative feedback.

THE GREENHOUSE EFFECT

People sometimes refer to the greenhouse effect as if it was a completely man-made phenomenon. In fact the greenhouse effect is perfectly natural. The basic mechanism involved is that, while the gases which make up the atmosphere are almost transparent to incoming short-wave solar radiation, some of them, making up only a small percentage of the atmosphere's volume, absorb some of the outgoing long-wave radiation emitted by the Earth's surface. These gases warm the atmosphere which in turn re-radiates heat, some of it back down to the surface. This additional heating forms the greenhouse effect (Fig. 3.2). Without it, the average surface temperature of the Earth would be about -18°C, a figure more comparable with Mars (-47°C) than the present 15°C, producing conditions in which life would be impossible. On our other neighbouring planet, Venus, in contrast, the greenhouse effect is much stronger, and surface temperature is around 477°C.

Concern over global warming relates to the fact that man's activities, particularly the burning of fossil fuels and land use changes, have been increasing the atmospheric concentration of various greenhouse gases.

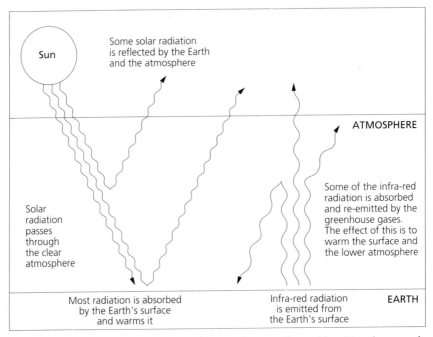

Fig. 3.2 Simplified diagram showing the greenhouse effect. (After Houghton et al., 1990)

not + man made gases.

Some of these gases added by man, such as carbon dioxide and methane, occur naturally. Others, like the highly radiation-absorbent CFCs, are completely man-made. The assumption is that, as the levels of these gases increase, average temperatures in the lower atmosphere and at the Earth's surface will rise. Global warming is thus caused by the enhancement of the natural greenhouse effect by man. The link between global warming and rising concentrations of greenhouse gases can be established by theoretical physics. In addition there is increasing evidence that in the past the sharp temperature shifts that were associated with the changes between glacial and interglacial conditions were accompanied by equally rapid alterations in atmospheric concentrations of carbon dioxide and methane (Chapellaz et al., 1993).

While orbital changes may have provided the background mechanism which began and ended ice ages there is strong evidence that greenhouse gas forcing amplified the warming and cooling trends. However, although it is accepted that there is a close link between greenhouse gas concentrations and average global temperatures, predicting the temperatures which are likely to result from specific rises in greenhouse gas levels is much more difficult. This is due to the sheer complexity of the climate system and the ways in which it may respond to greenhouse gas forcing.

Of the various trace gases in the atmosphere which control the green-house effect, water vapour is the most important, accounting for some 60–70 per cent of the natural greenhouse effect. However, as the water vapour content of the atmosphere is not directly affected by man's activities to a significant degree it will not be considered here. Carbon dioxide, which only accounts for about 25 per cent of the natural greenhouse effect, will be discussed in greater detail because its concentrations are much more susceptible to being increased through human activities.

The physical properties of the various greenhouse gases are well understood but calculating the impact on the greenhouse effect of increased volumes of them is more difficult. The contribution of a gas to the greenhouse effect depends on the wavelengths at which it absorbs infra-red radiation and the extent to which it overlaps with other gases. Gases absorbing at wavelengths similar to water vapour and carbon dioxide will contribute little to the greenhouse effect unless their concentrations are massive. There are, however, wavebands where absorption by carbon dioxide and water vapour is weak; a window region in which nearly 80 per cent of the radiation emitted in these wavebands escapes to space. Most of the other major greenhouse gases; methane, nitrous oxide, CFCs and tropospheric ozone, absorb in this window region. Some gases – methane and nitrous oxide for instance – overlap in terms of the radiation bands in which they absorb. This overlap reduces their heat-adding impact by about half. The man-made CFCs do not overlap with other greenhouse gases and this accentuates their effect. The problem is further complicated by major uncertainties regarding the natural and anthropogenic sources and sinks of some greenhouse gases; those for carbon dioxide and CFCs are better understood than methane and nitrous oxide.

It is one thing to calculate the direct radiative forcing effect on temperatures of given levels of increase in particular greenhouse gases, as their radiative properties are well understood. It is harder, however, to assess indirect effects resulting from chemical reactions within the atmosphere and feedback processes which may amplify or reduce the amount of warming. Forecasts of global warming also have to incorporate estimates of future emissions of the various greenhouse gases, estimates which may have quite wide margins of error.

As a result of their different radiative characteristics, comparable increases in different greenhouse gases have very different effects on radiative forcing. The impact of a gas depends not only on its concentration in the atmosphere but also on the strength of its radiation absorption by molecule or by weight. For example, in relation to carbon dioxide, methane has a radiative forcing about 21 times higher by molecule and 58 times higher by weight. CFC 12 has 16,000 times more radiative forcing than carbon dioxide per molecule but only 5,200 times by weight. Another influence on the impact of greenhouse gases is their atmospheric lifetimes. These vary considerably; methane, for example, is not very stable

Table 3.1 Direct global warming potentials for 20-year and 100-year-time periods

Gas	Direct GWP 20 years	Direct GWP 100 years	Lifetime (years)
Carbon dioxide	1	1	c120
Methane	35	11	10.5
Nitrous oxide	260	270	132
CFC 11	4500	3400	55
CFC 12	7100	7100	116
HCFC-22	4200	1600	15.8

(After Houghton et al., 1992)

and breaks down fairly quickly, while CFCs are very long-lived (Table 3.1). The representation of the impacts of different greenhouse gases according to various indices is a statistical minefield which has aroused vigorous debate (Caldeira and Casting, 1993).

Global warming potential (GWP) is a measure of the relative radiative effect of equal emissions of each greenhouse gas, taking into consideration the length of time they remain in the atmosphere. These are usually related to carbon dioxide (whose GWP is expressed as 1; see Table 3.1). Because of its more limited capacity to absorb radiation, carbon dioxide is the least effective greenhouse gas on a molecule-for-molecule basis. However, because vastly greater quantities of it are released by human activities compared to other greenhouse gases, it has been the most important man-enhanced greenhouse gas in the past, and is likely to remain so in the future. In the 1980s it accounted for 55 per cent of the total radiative forcing caused by additional greenhouse gases compared to 25 per cent for CFCs, 15 per cent for methane and 5 per cent for nitrous oxide (Fig. 3.3). On a longer time-scale, over the last 200 years the contribution of carbon dioxide to increased radiative forcing has been 61 per cent, methane 17 per cent, nitrous oxide 4 per cent and the recently-developed CFCs only 12 per cent (Fig. 3.4).

Rather than assessing the impact of various greenhouse gases individually it is often easier to consider their total forcing effect by relating it to the most important greenhouse gas, carbon dioxide, as the equivalent CO_2 concentration. This is the amount of carbon dioxide which would produce an amount of radiative forcing equivalent to that generated by all the different greenhouse gases acting together. Against the pre-industrial level of carbon dioxide, total radiative forcing to date has been equivalent to a 50 per cent rise in carbon dioxide. Carbon dioxide has actually risen by only about 26 per cent: the rest is the additional effect of the other greenhouse gases.

Because of the complexities of the atmosphere and the other environmental systems with which it interacts, especially the oceans which warm

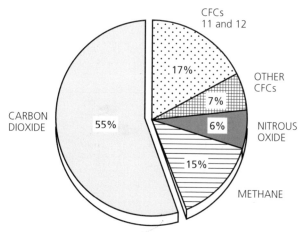

Fig. 3.3 Contribution of various greenhouse gases to the change in radiative forcing 1980–90. The contribution of ozone, which cannot be accurately assessed, is omitted. (After Houghton et al., 1990)

much more slowly, there is a time lag between a given increase in the concentration of greenhouse gases in the atmosphere and the achievement of the full or **equilibrium** warming which this increase generates. The amount of warming which has occurred at any particular time between an increase in greenhouse gases and the attainment of equilibrium warming is known as the realised warming. For example, calculations suggest that the increases of greenhouse gases which have occurred in the last 150 years due to man's activities have produced a rise in global mean temperatures of between 0.3 and 0.6°C. This is, however, only the realised warming. Due to the time lag there may be another 0.4–1.2°C of additional or committed warming still to come before equilibrium warming is reached, even if there were to be no further increase in greenhouse gases.

In terms of the activities which have contributed to the man-enhanced greenhouse effect, the burning of fossil fuels has accounted for an estimated 57 per cent, agriculture 14 per cent, deforestation 9 per cent and various industrial processes and non-energy uses (such as the manufacture of CFCs) 20 per cent. Only about 27–35 per cent of the increase in greenhouse gases has been generated by the developing world. The remainder has come from the industrialised nations. The balance is set to change markedly in future decades though. In 1988 the less developed countries only contributed 15 per cent of the additional carbon dioxide released that year from the burning of fossil fuels, but their input has been rising by around 5 per cent per annum against 0.7 per cent for the developed nations. It has been calculated that, in terms of carbon

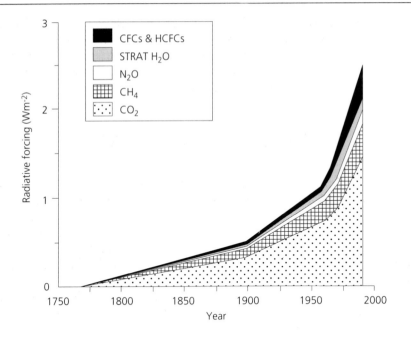

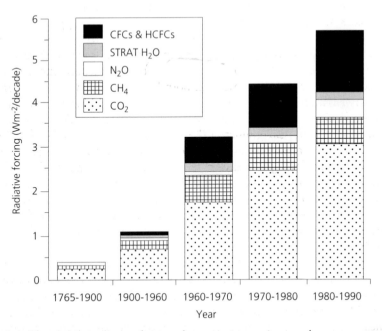

Fig. 3.4 Changes in radiative forcing due to increases in greenhouse gas concentrations between 1765 and 1990. (After Houghton et al., 1990)

dioxide emissions from fossil fuels, the underdeveloped world will catch up with the developed world by the middle of the next century.

Having considered the greenhouse effect and how its modification by human activity can be measured and assessed, it is possible to look in more detail at the major greenhouse gases.

CARBON DIOXIDE (CO_2)

Examination of ice cores has shown that carbon dioxide levels in the atmosphere have changed closely in step with temperature during the last 160,000 years. During interglacials, carbon dioxide concentrations were as high as 280 ppmv (parts per million volume), in glacial phases as low as 180 ppmv. During recent centuries down to around AD1800, however, carbon dioxide concentrations seem to have varied only within a range of about 10 ppmv around about 280 ppmv.

Since the early nineteenth century, carbon dioxide levels have risen by around 25 per cent from around 280 ppmv to about 355, a figure which has not been reached in the last 160,000 years at least. The annual rate of growth of carbon dioxide has been around 0.45 per cent or 1.5 ppmv. This is much faster than recorded natural rates of change in the past which only seem to have reached a maximum of about 0.5 ppmv per year. Atmospheric carbon dioxide concentrations have only been measured directly since 1958, first at Mauna Loa in Hawaii, a location distant from major local sources of human-generated carbon dioxide, and more recently in other suitably remote locations in both hemispheres. At Mauna Loa the level has risen from 315 ppmv in 1958 (Fig. 3.5). However, analyses of bubbles of air in Antarctic ice cores show that carbon dioxide began to rise slowly from the early nineteenth century, and more rapidly after 1900. The initial rise was due mainly to deforestation and the clearing of continental interiors for agriculture; especially in North America but also in South Africa, Australia, parts of South America and Russia (Fig. 3.6). In recent decades, however, the release of carbon dioxide from fossil fuels has been much more important. Carbon dioxide emissions from fossil fuel combustion overtook those from land clearance about the time of the First World War.

There are marked seasonal fluctuations in the amount of carbon dioxide in the atmosphere due to its intake and release by land plants through the processes of photosynthesis and respiration. These involve exchanges of about 50 Gt carbon a year (Gt = gigatonnes: 1 Gt carbon = 1 billion metric tonnes) with the atmosphere. Carbon dioxide concentrations increase in autumn and winter and fall in spring and summer, following the cycle of growth and dieback of vegetation. On a more local scale the seasonal amplitude in carbon dioxide content is 6 ppmv at Mauna Loa but as much

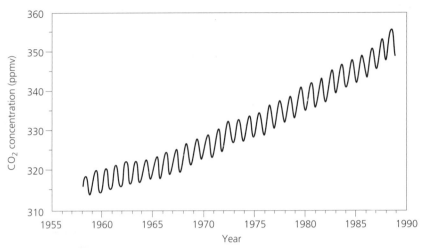

Fig. 3.5 Monthly average carbon dioxide concentrations at Mauna Loa, Hawaii. (After Wuebbles and Edmonds, 1991)

as 15 ppm at Point Barrow, Alaska due to the influence of land vegetation. Seasonal variations in carbon dioxide are less marked in the southern hemisphere due to the more limited extent of land. The amplitude of the seasonal cycle seems to be growing, which may reflect the increasing productivity of land ecosystems due to the carbon dioxide fertilisation effect (Chapter 6). Variation in global carbon dioxide levels from year to year can also be caused by ENSO events. When an event occurs, carbon dioxide concentrations tend to rise more than expected, possibly due to the reduced uptake of carbon dioxide by vegetation from the failure of the south-east Asian monsoon.

The overall trend of carbon dioxide in recent decades has been steadily upwards. In 1992–3, however, the growth rate of atmospheric carbon dioxide slowed by an amount unprecedented since detailed measurements began in 1958 (Sarmiento, 1993). From mid-1991 there was a sudden drop in carbon dioxide levels just after the Mount Pinatubo eruption. One suggested cause is that fallout from the eruption enriched areas of the ocean with iron which acted as a fertilizer and encouraged the productivity of phytoplankton, causing them to absorb additional carbon dioxide. Alternatively, the drop may have been linked with the terrestrial biosphere rather than the oceans. Did the cooling of temperatures due to the eruption affect the balance between photosynthesis and respiration? Despite this brief check, however, carbon dioxide concentrations seem set to continue their upward trend. The main uncertainty about the future is levels of emissions. There is a difference of ×3 between the highest and lowest of the IPCC emissions scenarios covering the period 1990–2100 (Siegenthaler and Sarmiento, 1993).

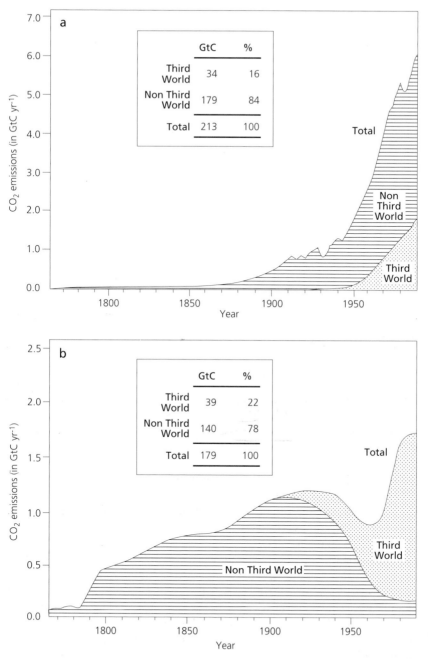

Fig. 3.6 (a) Fossil fuel-derived carbon dioxide emissions from Third World and non-Third World countries since 1765. (b) Carbon dioxide emissions from land use changes (mainly deforestation) for Third World and non-Third World countries since 1765. (After Mintzer, 1992)

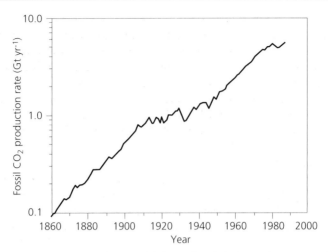

Fig. 3.7 Emissions of carbon dioxide from fossil fuel combustion and cement manufacture 1860–1990. (After Houghton et al., 1990)

We know more about the sources of carbon dioxide, especially emissions from fossil fuels, than other greenhouse gases and can quantify them with a smaller range of error (Fig. 3.7). Information on fossil fuel production allows the output of carbon dioxide from this source to be calculated for different countries, for the past as well as the present. At the moment, about 77 per cent of anthropogenic carbon dioxide comes from burning fossil fuels, the rest from deforestation and land use changes, though the increasing rate of tropical deforestation may alter this balance a little in the future.

About 5.4 Gt carbon are emitted each year from the burning of fossil fuels. Deforestation and agriculture account for around 1.6 Gt. This produces a total of about 7.0 Gt. These figures may sound precise but in reality they are only estimates with fairly wide ranges of error. It is easier to calculate the amount of carbon dioxide produced from fossil fuel combustion than from deforestation. Of this, 7.0 Gt of carbon – about half – remains in the atmosphere. The rest is absorbed by sinks such as the oceans and land plants. Estimates of the amount of carbon dioxide taken up by the oceans have varied from 2.0 GtC a year to as low as 1.0 Gt leaving around 1.6–1.8 Gt – about the same amount produced by deforestation – to be absorbed by the biosphere. Again there is a considerable margin of error and it is possible that the amount of carbon dioxide released from deforestation in tropical areas has been overestimated.

Between 1940 and 1988, around 35 Gt of carbon 'went missing' in that it could not be accounted for by the best estimates of uptake by the oceans and land plants (Mackenzie, 1993). It is now thought that the northern hemisphere's forests are a greater sink for carbon dioxide than had previously been realised. An analysis of tree ring data suggests that trees in northern and temperate forests have indeed been growing at faster rates for much of the twentieth century than previously, due to higher temperatures and possibly also to the carbon dioxide fertilization effect (Chapter 6). Prior to 1890 the northern forests were a net source of carbon dioxide due to felling and burning but, in the twentieth century, increased tree growth is thought to have outstripped the losses, so that the northern forests became a net sink for carbon until the 1970s when, due to increased felling, fires and the activities of pests, they may have become a net source of carbon again. Concern over tropical deforestation may in the future put more pressure on the northern forests; increasing timber output is expected in the future from many northern countries. It is also possible that some of the missing carbon may have been absorbed by deep-rooted grasses, some of African origin, which have been introduced into South American savannas (Fisher et al., 1994).

Land plants store almost as much carbon as the atmosphere; about 560 Gt, mainly in forests, against 725 in the atmosphere. This is estimated to be 15–20 per cent less than in the mid-nineteenth century. Soils hold about 1,300–1,400 Gt of carbon. Today, emissions from land use changes are dominated by tropical deforestation, accounting for between 0.4 and 1.6 Pg (Pg = Petagram: 1 Pg = 10^{15} grams) of carbon a year. The oceans store about 37,000 GtC. They absorb about 93 Gt a year and release only about 90 Gt. With higher temperatures there may be changes in the overall volume of carbon storage in the oceans and the level of carbon dioxide exchange which could add to or reduce atmospheric concentrations. Theoretically under the warmer conditions with greater precipitation which have been predicted (Chapter 4), the amount of carbon stored in vegetation and soils should increase. During the mid-Holocene climatic optimum (Chapter 2) there was certainly a much greater spread of forests in the northern hemisphere while grasslands covered extensive areas, like the Sahara, which are now arid. It has been estimated that approximately 100 Gt of carbon may have been lost from Africa alone since this time. This suggests that vegetation and soils should act as negative feedbacks on atmospheric carbon dioxide, slowing the rate of global warming but this news seems too good to be entirely convincing. It is not clear how fast carbon storage in vegetation and soils would respond to changed climatic conditions. The time lag in the response of vegetation in some areas, as discussed in Chapter 6, could be decades or even centuries. On the other hand, the increased carbon dioxide in the atmosphere should encourage plants to take up more carbon, redressing the balance to some degree.

Future levels of carbon dioxide in the atmosphere depend on a number of feedback mechanisms which are hard to quantify. Higher temperatures should increase photosynthesis in plants but may increase respiration even more, causing a net release of carbon dioxide. Higher temperatures may cause stress on ecosystems, causing additional release of carbon dioxide if plants cannot adapt sufficiently fast to changing temperatures and moisture conditions and die. Warmer conditions in higher latitudes may cause the release of more carbon dioxide as peat bogs dry out and their surface layers oxidize. Recent warming in the Arctic may already have begun to cause areas of the tundra to dry out, accelerating soil decomposition and the release of carbon dioxide, turning the tundra from a net carbon sink into a source (Webb and Overpeck, 1993). The warming of the oceans might reduce their capacity to absorb carbon dioxide from the atmosphere. Rises in ocean temperatures may also cause changes in ocean circulation patterns which may slow down the vertical mixing of ocean waters and, consequently, their ability to take up carbon dioxide.

All these influences would tend to increase carbon dioxide levels still further. On the other hand, higher concentrations in the atmosphere should increase the productivity of many ecosystems and crops causing enhanced plant growth which would absorb more carbon from the atmosphere. Equally the increasing use of artificial fertilizers and the stimulation of plant growth which occurs when their nitrates and phosphates are washed into natural ecosystems, would tend to cause more carbon dioxide to be absorbed by plants, though the impact of acid rain on vegetation would have the opposite effect. Ozone depletion, causing an increase in ultraviolet radiation at the Earth's surface, may also have detrimental effects on plant growth, releasing more carbon dioxide. Quantifying all these feedbacks and balancing them out is still impossible.

The production of carbon dioxide from the burning of fossil fuels has shown an exponential increase since the mid-nineteenth century with rises of up to 4 per cent per annum, interrupted by slower growth during the Depression of the 1930s, the two World Wars, and the aftermath of the oil price rises of 1973. In the early 1980s there was a cessation of growth in carbon dioxide output from fossil fuels, temporarily stabilizing it at around 5.3 Gt of carbon a year, but in the later 1980s it began to rise again reaching 6.0 GtC in 1990. There have, however, been important changes in the pattern of release of carbon dioxide from fossil fuels in recent years. The rate of increase from western Europe and North America fell from 3 per cent per annum between 1945–72 to around 1 per cent between 1973–84. The US accounted for 40 per cent of global fossil fuel carbon dioxide emissions in 1950 but under 25 per cent in 1990. Output from the USA, UK, Germany and Japan has grown little since the 1970s. In the USA, emissions peaked in 1979 at 1.3 Pg of carbon a year and were still lower than this in 1986. At the same time the rate of increase from developing countries has risen to 6 per cent per annum. When the growth of global fossil fuel emissions was 4 per cent per year it was projected that carbon dioxide concentrations would reach 600 ppmv early in the next century. With the rate of growth of fossil fuel emissions expected to be lower than this in future, the 600 ppmv level will not be reached until much later in the century.

The global warming potentials of other greenhouse gases are usually calculated in relation to the climate forcing effect over time of a unit of carbon dioxide. It is important to appreciate that the high GWPs of some greenhouse gases do not necessarily mean that they are a more serious problem than carbon dioxide. Because of its long atmospheric lifetime – 20 per cent of the carbon dioxide emitted today will still be around 500 years from now – climate forcing from increased carbon dioxide will remain the core of the global warming problem especially because, as will be discussed in Chapter 7, emissions of carbon dioxide are likely to be harder to control than those of some other greenhouse gases.

METHANE (CH$_4$)

Methane is less important as a greenhouse gas than carbon dioxide but it is over 21 times as effective, molecule for molecule, and its concentration has been increasing twice as fast. Analysis of air bubbles from Antarctic ice cores has shown that, in the past, methane concentrations have varied markedly in line with temperature, not only during glacial/interglacial transitions but during shorter climatic phases such as the Younger Dryas and the Little Ice Age (Chapellaz et al.,1993).

This suggests the existence of a strong feedback between the biosphere and climate on a time-scale of centuries. Minimum concentrations of methane during recent ice advances were around 350 ppbv, rising rapidly to 650–700 in the early stages of interglacials. Explanations for this increase have included the release of methane from the exposure of high-latitude peat bogs with melting ice and permafrost. However, the detailed chronology suggests that the rise in methane concentrations at the end of the last glaciation occurred when the American and Scandinavian ice sheets were still extensive, so that this suggestion does not seem entirely likely. Attention is now turning to tropical wetlands as potential methane sources at this time, especially the possibility that greatly enlarged lakes in the Sahara, created during wet spells before and after the Younger Dryas cold phase (Chapter 2), caused major methane releases around 14,000 and 11,500 years ago (Street-Perrot, 1993).

The pre-industrial concentration of methane in the eighteenth century is thought to have been around 700 ppbv. By 1988 this had more than doubled to 1720 ppbv with growth rates of around 1 per cent per annum (Fig. 3.8). Its contribution to the man-enhanced greenhouse effect may be as much as 17 per cent or even 20 per cent but its atmospheric concentration is less than 1 per cent of that of carbon dioxide. Nevertheless, if present trends were to continue, methane could be as important as carbon dioxide in terms of greenhouse forcing by the middle of the next century. Unlike other greenhouse gases, however, the rate of increase of methane

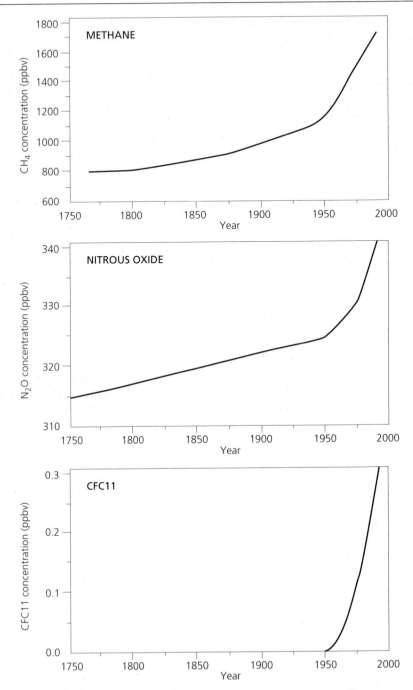

Fig. 3.8 Changes in the atmospheric concentrations of methane, nitrous oxide and CFC11. (After Houghton et al., 1990)

has slowed down markedly in recent years from around 20 ppbv a year in the late 1970s to around 10 ppbm in the early 1990s. The reasons for this are not clear, whether due to reduced emission or increased removal. The greatest falls have been in the mid- to high latitudes of the northern hemisphere, suggesting that the reason lies in these regions. Tighter controls over wastage and the consequent release of methane from natural gas production in Russia and eastern Europe are suspected as a major cause.

Methane is produced by the activity of bacteria breaking down organic matter in relatively oxygen-free conditions. Natural releases currently account for about 30 per cent of the total. Recent work has shown that levels of methane emissions are closely related to net ecosystem productivity, a variable which integrates many of the factors controlling methane emissions from wetlands, and is relatively easy to measure (Whiting and Chantonn, 1993). Anthropogenic methane sources are related to fossil fuel production and agriculture, especially paddy rice cultivation and livestock numbers, as well as to the burning of biomass, so that the recent increase in methane levels has close parallels with the growth of human population rather than following the rate of economic development (Table 3.2). Methane is also released in the extraction of coal and natural gas. Annual emissions are between 400(350) and 600 Tg (Tg = Teragrams: 1 Tg = 10^{12} grams) Most of this is destroyed in the atmosphere but about 40–50 Tg accumulate each year.

Sources of methane are more varied and harder to quantify than those of carbon dioxide. Indeed it is by no means certain that all the sources have been identified. As Table 3.2 shows, the most significant sources of

Table 3.2 Estimated sources of methane

	%	Tg per year
Natural		
Wetlands	21	115
Termites	4	20
Oceans	2	10
Freshwater	1	5
Hydrates	1	5
Anthropogenic		
Coal mining, natural gas & petroleum	19	100
Rice paddies	12	60
Fermentation in ruminant animals	16	80
Animal wastes	5	25
Domestic sewage treatment	5	25
Landfills	6	30
Biomass burning	8	40

(After Houghton et al., 1992)

methane are natural bogs and wetlands, rice paddies and cattle followed by natural gas, biomass burning and industrial processes. Recent work has suggested that the IPCC estimate for release from rice paddies may be too high. Emissions are hard to quantify since they vary considerably according to how individual rice paddy areas are managed. Figures quoted in the IPCC report (Houghton et al., 1992) from various specific rice paddies in different parts of the world range from 5–170 grammes per square metre per year. Ruminants – cattle and sheep – require bacteria to digest the cellulose that they consume and, in the process, large quantities of methane are produced. Emissions from cattle vary with the size of cow. A carefully-bred cow from a developed country produces about 55 kg a year, a lean one in an underdeveloped country about 35 kg, a sheep 5–8 kg. In terms of total emissions from ruminants, cattle are by far the most important, accounting for three quarters of ruminant methane. World cattle numbers have risen greatly over the last 150 years. Their impact may have been partly balanced by reductions in the numbers of wild ruminants such as bison, but overall there has probably been a significant net increase in the release of methane from such sources.

The main way in which methane is removed operates within the atmosphere itself. Methane is a relatively unstable and short-lived gas with a lifespan of only about eleven years. It reacts with hydroxyl (OH) radicals, the so-called 'detergent of the atmosphere'. About 90 per cent of the annual output of methane is destroyed in this way. However, in the process methane is oxidized to carbon dioxide, adding indirectly to global warming. Some methane reaches the stratosphere where its oxidation is a source of stratospheric water vapour which also acts as a greenhouse gas. Methane concentrations, clearly, are influenced by the amount of hydroxyl available for breaking down the gas. The amount of hydroxyl in the atmosphere depends partly on the amount of carbon monoxide (CO) which, along with methane and other hydrocarbons, reacts with the hydroxyl radicals. But carbon monoxide is itself produced by the breakdown of methane. Methane is also involved in the formation of tropospheric ozone, another greenhouse gas. The ozone in turn competes with methane for reaction with hydroxyl, further amplifying the increase of methane.

Future emissions of methane may contribute between 10 and 20 per cent of radiative forcing but, although the rise in methane concentrations has been slowing down in recent years, there are fears that this may be only a temporary respite. Warmer conditions are likely to increase the net primary productivity of wetland ecosystems and, in the process, methane emissions (Dacey et al., 1994). Even more alarming is the possibility that global warming may trigger off massive releases of methane from sources which are currently almost inactive, greatly amplifying the warming trend. Large quantities of methane are sealed in northern peat bogs by permafrost. If high-latitude areas warm more than the global average in decades to come, as many scientists think likely, the melting of permafrost could release large quantities of methane with the exposure of peat and enhanced decomposition of its organic matter. Huge amounts of methane are also sealed in ice-bound sediments under permafrost and ice sheets and in sea-floor sediments in arctic areas. The ability of these to react to greenhouse warming is uncertain but the amount of permafrost melting needed to expose methane from this layer would be great and changes might not be significant for several decades. Nevertheless, it has been suggested that the massive release of methane from such sources at times when Milankovitch cycles produced a radiation maximum in high latitudes may have been a major force in ending ice ages. So the response of existing undisturbed arctic reservoirs of methane could be substantial.

Methane emissions from energy production could certainly be reduced, as could seepage from landfills. Such methane can be burnt as an energy source. Different systems of rice paddy management might cut down the release of methane from this source considerably but very little detailed information is available on methane emissions from particular rice paddies under specific sets of conditions. Much more research is needed to determine how rice paddy management could be improved, but in any case such changes could be difficult to implement.

NITROUS OXIDE (N₂O)

Nitrous oxide is generated mainly from soils, especially by the use of fertilizers and the taking in of land for agriculture (Table 3.3). About 90 per cent of global emissions are thought to come from soils. The huge rise in the production of synthetic fertilizers in recent decades has increased the release of nitrous oxide to around 100 Tg of nitrogen a year. The pre-industrial level of around 280 ppbv, which had been stable for about 2,000 years, began to rise from the early nineteenth century and now stands at about 310 ppbv, a rise of around 10 per cent, currently increasing by about 0.3 per cent per annum (Fig. 3.8). This represents a higher concentration than at any time within the last 45,000 years (Levenberger and Siegenthaler, 1992).

As with methane, there are great gaps in our knowledge of the major sources and sinks of nitrous oxide and the amounts involved as the wide range of estimates in Table 3.3 shows. Only a few years ago it was thought that the major source of nitrous oxide was the burning of fossil fuels and biomass. This is now thought to be only a minor contributor. Recently identified sources include the production of nylon and nitric acid. Nitrous oxide is fairly stable in the atmosphere with a lifetime of about 150 years. Its potential as a greenhouse gas is about 230 times that of carbon dioxide molecule for molecule, so although its concentration is far smaller it is a potentially significant contributor to global warming, having accounted

Table 3.3　Estimated sources of nitrous oxide

	Tg of nitrogen per year
Natural	
Oceans	1.4 – 2.6
Tropical soils	2.7 – 5.7
Temperate soils	0.05 – 2.0
Anthropogenic	
Cultivated soils	0.03 – 3.0
Biomass burning	0.2 – 1.0
Stationary combustion	0.1 – 0.3
Mobile sources	0.2 – 0.6
Acid production	0.5 – 0.9

(Source IPCC 1992)

for about 5 per cent of the cumulative radiative forcing to date. Faster cycling of nitrogen in a warmer, wetter world might lead to increased emissions. Nitrous oxide is broken down in the stratosphere by ultraviolet radiation in photochemical reactions which contribute to the depletion of ozone through the production of nitric oxide (NO).

Conversion of rain forest to pasture is an important source of nitrous oxide. Research has shown that there is a ×5–×8 increase in release from the soils of recently cleared pastures in the Amazon compared with the original rain forest soils. The conversion of forest to pasture may supply up to 25 per cent of the current increase in nitrous oxide. However, release of nitrous oxide from such land only lasts for a few years, after which the amount falls off to half or one third of those of the original forest soils (Keller et al., 1993). Failure to appreciate this in the past has probably led to overestimation of the contribution of deforestation to nitrous oxide increases.

Forecasts for future emissions of nitrous oxide vary; growth rates to AD2050 have been estimated as between 0.5 and 1.75 per cent per annum. Under such scenarios the gas is likely to play a relatively modest role in future radiative forcing, perhaps about 5 per cent.

CHLOROFLUOROCARBONS (CFCS)

After carbon dioxide, chloroflurocarbons (CFCs) have attracted most attention as greenhouse gases partly because of their association with ozone depletion. A purely man-made product, their sources as aerosol spray propellants, in foam packaging, cleaning solvents and as refrigerants are well known. Their recent rate of increase in the atmosphere has been up to 5–5.5 per cent per annum, faster than any other greenhouse gas

(Fig. 3.8). The concentrations of the two main gases, CFC11 and CFC12, are 280 and 480 pptv (parts per trillion volume) respectively. These gases have relatively long atmospheric lifetimes; 55 years in the case of CFC11, 116 years for CFC12. In the 1980s they contributed about one third of the radiative forcing of gases other than carbon dioxide. Molecule for molecule these are tremendously powerful greenhouse gases, several thousand times more effective than carbon dioxide. They were first produced as coolants in the 1930s. From the 1950s their concentration rose fast, as CFCs leaked from disused fridges and air conditioning systems. During the first half of the 1980s they were thought to have been responsible for 20 per cent of the total radiative forcing due to greenhouse gas increases. The Montreal Protocol and later agreements have bound most countries to cut production drastically and phase out CFCs entirely by AD2000. The Montreal Protocol reflected the desire to protect stratospheric ozone from CFCs. Production of these gases was to be frozen at 1986 levels by 1 July 1989, 80 per cent of these levels by 1 July 1993 and 50 per cent by 1 July 1998. The 1990 London Agreement stipulated that production of CFCs should end completely by AD2000.

Between 1986 and 1992 consumption of the main CFCs in developed countries dropped by around 50 per cent, a figure representing an improvement on the Montreal agreement. Growth rates of atmospheric concentrations from 1977–84 were linear; CFC11 9 pptv/year and CFC12 17 pptv/year. By 1985–88 this had risen to 11 pptv and 19.5 respectively. Now it has dropped to 2.7 and 10.5 (Elkins et al., 1993). The levelling off since 1989 has been due to reduced production, especially of aerosol propellants. However, because of their high global warming potential and their long residence time in the atmosphere these gases will continue to contribute to the enhanced greenhouse effect for a century or so after production has ceased. They are being replaced with halogenated hydrocarbons, some containing chlorine (HCFCs) others not (HFCs). These have a shorter life than CFCs; generally less than 20 years, and are broken down in the lower atmosphere by reaction with hydroxyl radicals. Their atmospheric concentrations will thus be lower than those of CFCs, but although they do not deplete ozone these are still greenhouse gases and could affect climate if their concentrations become sufficiently large. Nevertheless, it has been calculated that, if the Montreal Protocol is adhered to, radiative forcing over the next few decades will be reduced and the rate of global warming will be slowed down.

OZONE

High-level ozone is a major absorber of short-wave ultraviolet radiation and is of primary importance in maintaining the temperature structure of the stratosphere. Ozone at lower levels is also an important absorber of long-wave infra-red radiation so that it is also a greenhouse gas. Below about 30 km the addition of ozone causes net atmospheric warming and above this height, cooling. Stratospheric ozone accounts for about 90 per cent of the atmosphere's concentration. About 10 per cent of atmospheric ozone occurs lower down in the troposphere where it acts as a greenhouse gas. It is produced in photochemical reactions involving gases like methane, other hydrocarbons, carbon monoxide and nitrogen oxides. These are all formed during combustion of fossil fuels. Our knowledge of its effect is limited by lack of data but some estimates have placed its contribution to radiative forcing during the 1980s as high as 8 per cent. Tropospheric ozone breaks down under ultraviolet radiation into oxygen atoms which react with water vapour to produce hydroxyl, which helps regulate levels of methane. Tropospheric ozone has a lifetime of only a few weeks at most and its levels have only been monitored since the 1970s. Highest concentrations occur over industrialised areas. Amounts in Europe seem to have increased at around 1–2 per cent per annum and are over twice as high now as they were before the 1950s.

FEEDBACK MECHANISMS AND THE GREENHOUSE EFFECT

These are the main greenhouse gases whose concentrations have been affected by man. It is worth noting, however, that atmospheric water vapour, the most important natural greenhouse gas, may increase in a warmer world with greater evaporation, enhancing global warming still further. The oxidation of methane in the stratosphere also produces water vapour which may have similar effects. A number of other feedback mechanisms may come into play to enhance and, in some cases, damp down the effects of rises in greenhouse gas levels (Table 3.4). As atmospheric temperatures increase, the surface layer of the ocean will warm and may release additional carbon dioxide in solution into the atmosphere. On the other hand, the depletion of stratospheric ozone as a result of reactions involving CFCs will lead to reduced absorption of incoming radiation, and a cooling effect. Recent research suggests that this may be about equal in magnitude to the additional contribution of CFCs to greenhouse warming, effectively negating their impact. The balancing of these two influences may be one of the reasons why global temperatures have not risen as much in recent decades as predicted for the increase in greenhouse gases which has occurred.

Clouds form another source of feedback, though whether their net effect on global warming is likely to be positive or negative is still far

Table 3.4 Climatic feedback mechanisms and greenhouse forcing

Feedback	Direction
Clouds: high altitude	Positive
Clouds: low altitude	Negative
Clouds: overall effect	Positive or negative
Water vapour	Positive
Ice-albedo effect	Positive
Aerosols	Negative

(Source IPCC 1992)

from clear. It is thought that, under present conditions, the net effect of clouds is a slight cooling one. As temperatures rise there should be more evaporation from the oceans and increased cloudiness. This could result in more radiation being reflected back, reducing temperatures. However, if warming of the atmosphere displaces clouds to higher, colder levels this may produce a positive feedback because the colder cloud will emit less radiation and so enhance the greenhouse effect. It is thought that low clouds become more reflective as temperature increases, causing negative feedback while feedback from high clouds under warmer conditions might be positive or negative. Higher temperatures at higher latitudes, with increased poleward transport of moisture, might produce more snow with a higher albedo which has the effect of reducing temperatures.

As we have seen, concentrations of greenhouse gases depend not only on levels of emissions but also on chemical processes operating in the atmosphere. Apart from carbon dioxide it is mainly chemical processes in the atmosphere that affect the rate of removal of greenhouse gases. The main removal mechanism for methane is reaction with hydroxyl radicals in the troposphere. Nitrous oxide and CFCs are destroyed by photodissociation (dissociation of molecules involving reactions with a solar photon). Hydroxyl is not a greenhouse gas but, because it is the main scavenger in the troposphere, its levels affect the concentrations of several greenhouse gases. Production of hydroxyl involves the reaction of oxygen atoms, produced by the **photolysis** (chemical decomposition under the action of light) of ozone, with water vapour. The higher levels of ozone or water vapour in the troposphere which may occur with warmer conditions should increase the amount of hydroxyl and accelerate the removal of greenhouse gases such as methane. Conversely, increases in methane or carbon monoxide should decrease the amount of hydroxyl, causing a positive feedback enhancing global warming.

Aerosols

In addition the existence in the atmosphere of aerosols – small particles of dust, salt, soil and sulphates from fossil fuel combustion – must be considered. Their effects on the atmosphere are complex and not well understood. At high levels in the atmosphere they tend to cause cooling but at lower levels in the troposphere they may enhance the greenhouse effect. About half the aerosol particles in the atmosphere are natural in origin. Most natural aerosol particles come from the ocean; a biological waste product, dimethyl sulphide, is emitted from the ocean surface and is oxidized in the atmosphere to form sulphate aerosols. Most of today's sulphate aerosols come, however, from the burning of fossil fuels, destruction of tropical rain forests and other human activities. Sulphate particles are the main source of nucleii around which water vapour condenses, producing clouds. Because their lifespan in the atmosphere is only days or weeks there are substantial regional variations in their atmospheric concentrations.

Aerosols have a direct effect in blocking incoming solar radiation. They also have important indirect effects. They act as nucleii on which cloud droplets condense. For a given amount of water vapour, more aerosols produce more, smaller droplets. These are more reflective than fewer, larger ones so the albedo of clouds is increased (Wigley, 1994). They also increase the vertical and horizontal extent of clouds (Pincus and Baker, 1994).

The 1992 update to the IPCC report (Houghton et al., 1992) suggested that sulphate particles tend to have an overall net cooling effect on global climate. They cool the Earth directly by reflecting incoming solar radiation and indirectly by generating increased cloud cover. Indeed, it has been suggested that if it had not been for the increase in emissions of sulphate particles during the last century the degree of global warming might have been twice the 0.5°C actually observed. The pall of pollution from northern industrial nations which has spread over the Arctic has reduced the amount of solar radiation reaching the surface by 15 per cent. This may be the reason why this region has not warmed in recent years on the scale of the Antarctic (Stephens, 1994).

We have seen that the warming trend which has been evident in the last century has been more continuous in the southern hemisphere than the northern. This may be linked to greater release of sulphate particles from the industrial nations in the northern hemisphere producing more condensation nucleii and increased cloud cover, reflecting more solar radiation than in the southern hemisphere. There is a certain amount of observational data suggesting that cloud cover has indeed increased over areas such as eastern North America. The 1992 IPCC update suggested that sulphate particles may have had a significant effect in reducing the rate of warming in the northern hemisphere which would have been expected from the build-up of greenhouse gases to date. This effect may have been sufficient to cancel out the impact of increased carbon dioxide or, alternatively, of two other greenhouse gases combined. A study by the Meteorological Office has suggested that sulphate particles may reduce the rate of future warming to as little as 0.15°C per decade, half the rate estimated by the IPCC.

Aerosols should not, however, be seen as merely cancelling out the effects of greenhouse gas warming. By pulling the atmosphere in an opposite direction to greenhouse gas forcing, and doing it at a regional rather than global scale, they are generating stresses, the climatic implications of which are uncertain (Taylor and Penner, 1994). Despite the global rise in average temperatures which has occurred in the last century some areas with high sulphur dioxide emissions may actually have cooled, while the discovery that the observed temperature rises have occurred mainly at night has been reported from areas as far apart as the eastern USA and China. If the production of aerosols increases, it may lead to cooling in middle and high latitudes of the northern hemisphere which could increase the temperature gradient between these regions and the tropics, producing a faster circulation and more storminess in mid-latitudes. There are indications that storminess may already have increased in recent years in the North Atlantic area.

This discussion of the feedback mechanisms which complicate the study of the effects of rising greenhouse gas concentrations on climate, highlights the complexity of the climate system and emphasises the difficulties involved in predicting its behaviour. But how can climatologists forecast what may happen to climate on global and regional scales if the build-up of greenhouse gases continues? This is the topic turned to in the next chapter.

|4|

Predicting Future Climates

In Chapter 1 we saw something of the variety of evidence that can be used to reconstruct past climates on time-scales from decades to hundreds of thousands of years. We know more and more about what is happening in the atmosphere today from increasingly sophisticated techniques which range from surface observations to satellite monitoring. But how do we know what the climate will be like in the future? It is vital to understand how climatologists can construct scenarios for future climates and to realise the limitations of the techniques they use to appreciate just how much or how little confidence can be placed in their assertions. Equipped with this knowledge it is possible to take forecasts of temperature rises resulting from greenhouse forcing and relate them to the techniques and data which have been used and also to the assumptions which have been made about future changes in greenhouse gas concentrations.

The problems involved in predicting future greenhouse gas emissions and how climate will respond to them in turn explain why there is so much uncertainty regarding the scale of future global warming and how rises in global average temperatures will be translated into climatic changes in particular regions. Projections of future climate build in ranges of uncertainty because of the difficulty of estimating future greenhouse gas concentrations, depending on what measures may be taken to reduce emissions, the simplifications involved in the techniques themselves, the limitations of the data they use, and uncertainties about how the climate system operates.

The consensus among scientists is that the concentrations of various greenhouse gases in the atmosphere have increased during the last 200 years due to human activity, and that, unless action is taken, the increase will continue. Higher concentrations of greenhouse gases will affect the Earth's energy balance and alter the radiative forcing of the climate system, enhancing the greenhouse effect. As a result the Earth's surface and lower atmosphere will, on average, become warmer. How much

warmer will depend on how sensitive the climate system is to such radiative forcing changes as well as on other influences such as the slowness
with which the oceans absorb heat from the atmosphere. Greenhouse gas
forcing may be the cause of the warming which has occurred during the
last century. Average global surface air temperatures have risen about
0.5°C during the last century with a sharp rise in the 1980s. The scale of
this warming is in line with the predicted climatic response to the increases
in greenhouse gases which has occurred. However, it also lies within the
natural limits of variability of the climate system, so that it cannot yet be
proved conclusively that greenhouse gas increases have caused this recent
global warming. Because several greenhouse gases – such as carbon
dioxide, nitrous oxide and CFCs – have very long atmospheric lifetimes,
it will take a long time for any policies aimed at reducing emissions to
begin to slow or halt the process of global warming. Continued emissions
of these gases at current rates will result in increased atmospheric concentrations for centuries (Houghton et al., 1990).

A rise in all greenhouse gases, equivalent to a doubling of the pre-industrial level of carbon dioxide, is likely to be reached by the middle of the
next century. Various studies have suggested that a doubling of
atmospheric carbon dioxide, or its equivalent incorporating the effects of
increases in other greenhouse gases, may raise mean global temperatures
between 1.5 and 4.5°C, with a best guess at around 2.5–3.5°C.

The longer emissions of greenhouse gases continue to rise, the more
drastic will need to be the cutback to stabilise these gases at levels which
will eventually halt the process of global warming. The IPCC predicted,
on the basis of its 'business as usual' computer climate modelling scenario,
that future global average temperatures were likely to rise at around 0.3°C
per decade producing an increase of around 1°C by 2025 and 3°C by
2100, though more recent studies have revised these estimates downwards.
It is important to appreciate that these are realised temperature rises, not
committed ones. Even if emissions increases were to be halted, greenhouse
gases already in the atmosphere should continue to deliver further temperature rises for many decades before equilibrium conditions are eventually
reached. Even this fairly conservative forecast of possible temperature
changes would, if it occurred, be both significant in its effects and much
faster than the warming which has taken place in the last century.

The last time the world was 2°C warmer than at present was during
the last interglacial period, some 120,000 years ago. The projected rate of
temperature change would be faster than any which are known to have
occurred naturally with the possible exception of the most rapid rates of
warming at the end of the last glaciation (Chapter 2). These predictions
are, however, only global averages. Climatic changes on a regional scale
would be very varied in ways which are much harder to predict. Some
areas might even become colder while others could warm up considerably
more than the global average.

These predictions acknowledge the limitations of our current knowledge. Future research might reveal sets of interactions which could damp down the man-enhanced greenhouse effect and reduce future warming. Equally, there might be some unpleasant surprises in terms of positive feedback mechanisms which accentuated the warming trend. The discovery in 1984 of the hole in the ozone layer shows that such surprises might be of major significance. The idea that, on the basis of current knowledge, we may be underestimating the degree to which built-in mechanisms might substantially reinforce the predicted level of warming should not be seen as a media scare story, but as a real possibility. On the other hand, it is possible that elements of inertia, especially the time lag due to the slowness of the oceans in taking up heat from the atmosphere, may mean that the time-scales of change suggested above are too short and that changes may occur more slowly, giving more leeway for human society to adapt to new conditions. Nevertheless, climatic changes on the scale predicted would have an undoubted impact on natural ecosystems, agriculture, and many other aspects of human activity as well as causing a rise in mean sea levels.

The two main methods of producing future climate scenarios are the study of analogues drawn from past periods when global climate was warmer than now and the use of complex computer models which simulate the behaviour of the atmosphere.

ANALOGUES FROM PAST CLIMATES

The assumption behind the use of analogues is that, given similar background conditions, climate responds in a similar way to different forcing factors, so that as long as features such as the extent of land and sea and the pattern of ocean circulation remain the same, it makes no difference whether warming is due to changes in the concentration of greenhouse gases or in solar radiation.

Analogues can be taken from recent periods covered by detailed instrumental records, from earlier historic times or from the more distant past when only geological evidence of past climates is available. For early periods, such as the last glacial maximum or the postglacial climatic optimum, geographical coverage of data at a global scale may be very uneven. It may also be impossible, from the nature of the evidence available, to reconstruct many key climatic variables. The use of recent instrumental records gets round this problem. Runs of unusually warm years can be compared with longer-term averages to establish what climate might be like in the future under warmer conditions. The difficulty here is that, within the period covered by widespread sets of meteorological data, the range of climatic variability has been small and only limited temperature differences are involved, so that using this type of analogue may only provide indications of possible climatic patterns in the early stages of global warming.

The problem with past periods of warmth, such as the postglacial climatic optimum, is whether they really are true analogues for a world warmed by anthropogenic enhancement of the greenhouse effect. Drawing analogues between past and future climates should only be done with caution as boundary conditions may not be comparable. One difficulty with analogues from the period beyond widespread instrumental records is that, whatever produced warmer conditions, it was certainly not an increase in the atmospheric concentration of greenhouse gases due to human activities. It is often suggested that the mid-Holocene optimum provides a useful analogue for future warmer conditions. The Holocene climatic optimum was about 2°C warmer than at present and might seem to offer a glimpse of what global climate might be like towards the middle of the next century if the rate of global warming proceeds roughly as predicted. However, forcing factors were not the same as they are today, because the Earth's orbital parameters were different. The higher temperatures experienced over much of Europe north of the Mediterranean involved a peak of summer warmth due to the Earth being closer to the Sun during the northern hemisphere summer rather than winter as it is today. This enhanced the contrasts between seasons as well as between continental interiors and coastal areas. The obliquity of the Earth's axis was greater than now and the planet's orbit was closer to the Sun in June rather than January as at present. As a result, the northern hemisphere received about 5 per cent more solar radiation in summer than it does now. Warming at this time was most evident in mid- and high latitudes in summer and may have been linked to circulation patterns which were significantly different from those of today. However, computer modelling of likely future climate suggests that warming is likely to be more pronounced in these latitudes in winter rather than summer.

Warmer climates in the past also seem to have been moister ones: in the last (Eemian) interglacial, when mean annual temperatures may have been 2–3°C above present levels, there are indications that precipitation may have been 30–50 per cent higher. Equally, the mid-Holocene optimum was moister over much of the northern hemisphere, including the Sahara (Chapter 2).

Past fluctuations in climate have indeed been associated with significant changes in greenhouse gas concentrations but not in the same mixture as at present, at the same levels, or against the same background conditions (Chapters 2 and 3). There is a close parallel between past temperature fluctuations and levels of carbon dioxide and methane. The concentrations of these gases were lower during glacial periods and higher during interglacials. Carbon dioxide levels rose by around 80 ppmv at the end of the last ice age and temperatures rose by about 7°C but this is not to say that an 80 ppmv rise in carbon dioxide was the prime cause of the temperature increase, or that such a rise in carbon dioxide concentrations in the future would be accompanied by a comparable change in temperature. In fact the rises in carbon dioxide and methane at the end of the last glaciation were less than those which have occurred during the last two centuries. So, despite the close correlation, only part of the changes associated with the glacial/interglacial cycle can

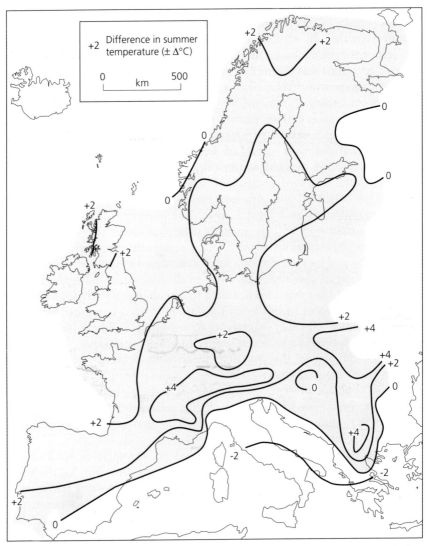

Fig. 4.1 Differences in summer temperatures in Europe about 6,000 years before present, against those of today. The stipple shows areas where temperatures in the mid-Holocene were higher than today. (After Roberts, 1994)

be attributed to greenhouse gas forcing. Computer modelling of the climate at the last glacial maximum suggests that the fall in carbon dioxide levels reduced temperature by only around 1°C while other influences such as albedo changes due to the spread of land and sea ice had perhaps three or four times the impact. Nor is the link between temperatures and carbon dioxide concentrations an invariable one. During the postglacial climatic optimum, carbon dioxide levels were lower than they have been during the past 2,000 years.

The impact of the Milankovitch orbital cycles greatly reduces the usefulness of comparisons with the climates of the mid-Holocene optimum and the last interglacial (Fig. 4.1). Earlier climates, in warm eras before the Pleistocene, occurred at times when other boundary conditions, such as the positions of the continents and major mountain ranges, were significantly different from today. The medieval optimum is sufficiently close to the present in time for boundary conditions to have been comparable but it is far from clear that warmer conditions then were truly global.

Various attempts have been made to examine the regional distribution of precipitation during the mid-Holocene optimum and to suggest this as a possible analogue for the future (Fig. 4.2). In one scenario, most of

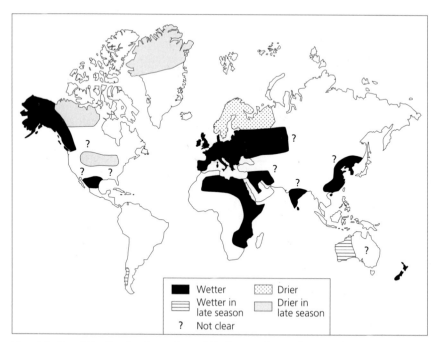

Fig. 4.2 Possible distribution of rainfall during the Holocene climatic optimum, relative to present levels. Blank areas reflect lack of information rather than no rainfall change. (After Henderson-Sellers and Robinson, 1986)

Europe comes out wetter, as does much of East Africa, the Middle East, India and parts of China, western Australia and New Zealand. On the other hand, the American Midwest and Great Plains as well as Scandinavia are forecast to be drier (Kellogg, 1978). Such reconstructions are useful for testing computer models of the atmosphere (see below), but as a forecasting method their use is limited. Studies of climatic conditions during more recent warm periods may provide more convincing pointers to the future (Pittock, 1983). They have the advantage over computer models of bringing out regional variations in considerable detail. A recent simulation of the effects of a warmer climate on the American Midwest and Great Plains used the climate of the Dust Bowl era as an analogue (Rosenberg, 1993).

Although dismissed as being of limited use by the IPCC report (Houghton et al., 1990), analogue climates have begun to be more fashionable again. The study of past climates emphasises some important general points for the future. For instance, the changes which have occurred in the areas of the world's deserts during the last glaciation and the Holocene, show the importance for man of changes in the distribution of moisture and indicate that very marked changes in such distributions can occur with only small shifts of temperature.

GENERAL CIRCULATION MODELS

The most widely used technique for constructing future climate scenarios is the use of complex simulations of the atmosphere called General Circulation Models (GCMs). These have been described as 'numerical laboratories' in which very complex experiments can be performed. They have also been described as 'dirty crystal balls' as far as foretelling the future is concerned! The first simple models were developed by the Climate Dynamics Group of the American Geophysical Fluid Dynamics Laboratory (GFDL) in the 1970s. During the 1980s other institutions developed their own models, including the Goddard Institute for Space Studies (GISS), the National Centre for Atmospheric Research (NCAR), Oregon State University (OSU), and the UK Meteorological Office (UKMO).

GCMs involve mathematical representations of climatic processes on three-dimensional grids of points at the Earth's surface and at a number of levels upwards through the troposphere and into the stratosphere. The horizontal resolution of the grid is often between 300–1,000 km and there may be a dozen or more layers. For each grid box a range of climatic data is input including temperature, precipitation, wind, humidity and pressure. The interactions between variables within the climate system are represented as accurately as knowledge and computer power allows. But models are, by definition, simplifications of reality. Because observational data are not sufficiently complete and accurate to represent real environmental conditions with 100 per cent accuracy there is, inevitably, an element of uncertainty from the very start in the results produced by these models.

Most of the models have been run by setting up two contrasting modes of climate. One represents conditions incorporating values for atmospheric carbon dioxide content which are equal to those of the present day or pre-industrial times. The model is then run through a series of simulated time steps for up to ten model years. Each time step allows the data in the model to interact for a particular length of time, often thirty minutes. The new sets of conditions resulting from these interactions then form the basis for the next time step, until the climate reaches an equilibrium state. The accuracy of a model can only be judged by comparing it with past or present climates to see how well it reproduces patterns in the real atmosphere. The accuracy of the model can then be verified by the degree to which it can simulate large-scale shifts in temperature.

A measure of the success of GCMs is their ability to simulate climatic patterns such as ENSO events, droughts in the Sahel, or the monsoon circulation. They can represent large-scale mid-latitude depressions but not smaller tropical cyclones. They are less good at simulating the complex behaviour of systems such as the upper westerly circulation. Their representation of real temperature patterns is considerably better than those of precipitation but in both cases, while the simulation may be fairly realistic on a global scale, representation of regional patterns is much more variable and less convincing.

The accuracy of GCMs can be tested in various ways. They can be compared with real climate on short time-scales to see, for example, how well they simulate seasonal variations. This is not as easy as it sounds because data from the grid points of GCMs have to be compared with those from an irregular distribution of meteorological stations. GCMs can be operated to see how well they simulate past climates such as those of the postglacial climatic optimum, the last glacial maximum, or the previous interglacial. They have also been used to demonstrate the impact of 'nuclear winter', the devastating changes in climate which would probably result from a major nuclear war. Different GCMs tend to contain errors which have a family resemblance due to the ways in which they are constructed: a tendency to exaggerate the coldness of the lower stratosphere in polar regions for instance. On the other hand, the fact that they can reproduce large-scale but short-lived features of the atmospheric circulation does not automatically mean that they can simulate the much smaller temperature variations associated with global warming, or that they are trustworthy when run over longer time periods.

For more recent times, GCMs can simulate a rise in global temperature resulting from the increase in greenhouse gas concentrations since the nineteenth century. On the basis of such models, global temperatures should have risen by about 1.0–1.8°C in this period. In fact, the real rise has been in the region of 0.5°C. This may indicate that not all of the complex feedback mechanisms involved have been taken into account. Once a model has been verified satisfactorily in this way it can then be run to produce scenarios for possible future climates.

VALIDATING GCMS

Once a model has been tested in this way it can then be run for a second time with all the parameters unchanged except that the concentration of carbon dioxide is increased – usually it is doubled. The new model is then run through simulated time phases until equilibrium conditions are reached. The different climatic patterns associated with the doubling of carbon dioxide can then be compared with the original control run and differences between them mapped to indicate how the equilibrium climatic conditions associated with an equivalent doubling of carbon dioxide might differ from those of the present. Overall, the ability of GCMs to simulate real climates to the level that they do gives climatologists reasonable confidence that they can reproduce some of the general features of climate in a greenhouse effect-warmed world.

There are, however, obvious problems with using GCMs. First, they involve simply doubling carbon dioxide levels while global warming, if it occurs, will be the result of forcing due to increases in a number of different greenhouse gases. The impact on climate may not necessarily be the same. Second, concentrations of carbon dioxide are doubled instantaneously rather than building up gradually, which is what is happening in reality. Such models do not provide indications of the rate at which warming would occur or the nature of the intermediate climates which would exist before an equilibrium climate for a particular level of carbon dioxide forcing was reached.

The models often contain major simplifications regarding the configuration of the continents: on many, the Mediterranean and Hudson Bay are treated as closed lakes while some omit the isthmus of Panama and Japan (Fig. 4.3). Topography is greatly simplified and only the main mountain ranges are shown. Output from models is generated for grid points and has to be interpolated for the grid squares between. The GCM variables most commonly plotted are temperature, precipitation, and soil moisture.

A major drawback of GCMs is that their horizontal resolution is poor, so that regions smaller than about 4 million km² cannot have their climates represented with any confidence. On this scale the British Isles, with its varied climate, might be represented by only two grid points. Higher horizontal resolution has been achieved in some models but at the expense of reducing the number of vertical layers. Only a limited number

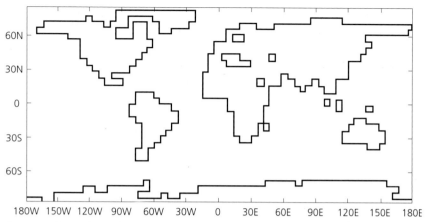

Fig. 4.3 The outline of the continents used in a typical GCM showing the coarseness of the resolution. (After Nilsson, 1992)

of higher-resolution models with adequate vertical representation of the atmosphere have been run. One might assume that those with the highest horizontal resolution would automatically provide the best representation of climate but this is not necessarily so.

A further problem of scale involves the ways in which GCMs represent many climatic processes. Many key aspects of climate, such as variations in cloudiness, operate at scales smaller than the GCM grids and so cannot be represented directly. To incorporate such elements parameterization is necessary. This involves averaging out the distribution of phenomena at grid square level using substitute data. To achieve this, climatologists look for relationships between a distribution which is too small to represent and one which can be modelled at grid scale, and use the latter instead; average humidity instead of cloud cover for instance. Because of the amount of simplification involved in parameterization, GCMs are less than convincing in their modelling of many important feedback processes.

Clouds are a major problem in GCMs because variations in their altitude, distribution and type have major impacts on whether they reflect or trap solar radiation. The ways in which they reflect or absorb radiation are not yet clearly understood. Much of the variation between GCMs is due to differences in the ways in which they handle the representation of clouds and their impact on radiation. At the moment, clouds are thought to have a net cooling effect on global temperatures but if, as some GCMs suggest, a warmer planet will generate more cloud cover at higher altitudes, the net effect may be to reinforce rather than offset global warming.

Soil moisture and the albedo of the Earth's surface are two other important influences which are represented rather crudely by GCMs. Soil is treated as a kind of shallow bucket which is filled by precipitation and emptied by evaporation with only one standard soil type. Neither are GCMs able to model accurately the complex interactions between the atmosphere and the oceans. Oceans have frequently been represented as a shallow slab only 70 m deep, incorporating only the upper layer and not taking into account exchanges between the surface layers and the depths of the oceans. The ice-albedo feedback is better understood and is handled more convincingly by GCMs. One advantage of GCMs is that it is possible to incorporate important but uncertain variables such as cloud feedbacks and vary them over a range of values to see how important they are in influencing climate sensitivity. Even if it is impossible to say which simulation is the most realistic it gives an idea of the range of climatic futures which could occur.

Equilibrium GCMs can only consider the effects on climate of changes in greenhouse gas concentrations, and not the impact of related factors such as deforestation and changes in cloud reflectivity due to sulphur emissions. Over twenty simulations using GCMs have been carried out by a number of research groups. Their results have many features in common but this is probably due in part to the fact that there are many similarities between the models in the ways in which they represent the processes operating in the atmosphere – potentially all of them could be wrong, or at least equally inaccurate.

It is important to remember that these models produce scenarios of what could occur which are not the same as predictions of what will happen. Within the limits of the models and the knowledge which has gone into constructing them, they represent what could happen under given sets of circumstances which might, or might not, occur in reality. They allow scientists to examine the relative importance of various elements, such as different greenhouse gases and aerosols, in influencing climate and help in our understanding of the relationships which cause climatic change. The results of scenarios may vary considerably from reality even over short time-scales. Even if conditions at a particular time in the future were identical to those built into a model, it still might not represent the resulting climate accurately because of its limitations or oversimplifications. This is especially true of climate at a regional scale.

Many GCMs still fail to reproduce features of the modern climate such as the strength and location of areas of persistent high and low pressure like the Azores high or the Icelandic low. Agreement between different models is especially poor in areas of sensitive climate such as north-west Europe where relatively small-scale changes in the atmosphere can have major impacts. For other areas, such as the American Midwest, there is greater agreement, but given that the models are constructed in the same basic fashion this does not necessarily mean that they are correct. Confidence is increased if the patterns shown by GCMs can be explained using known physical mechanisms, such as greater warming in high

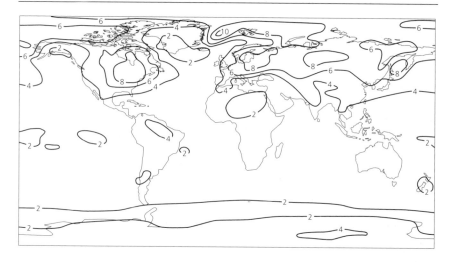

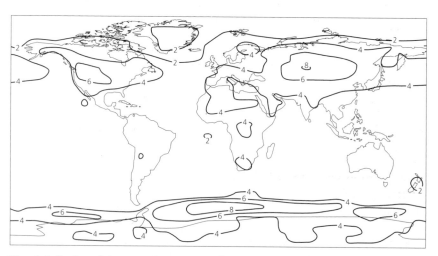

Fig. 4.4 Projected increase in mean surface air temperature (degrees C) with a doubling of CO_2 levels for the UK Meteorological Office (UKMO) model. Top: December, January, February. Bottom: June, July, August. (After Henderson-Sellers, 1993). Note the scale of increase in high northern latitudes during the northern hemisphere winter.

latitudes of the northern hemisphere being associated with the ice-albedo feedback mechanism (Fig. 4.4). Confidence is reduced once one moves from a global to a regional scale and from temperature to other climate variables such as precipitation (Fig. 4.5).

Transient climates, occurring in the early phases of global warming, may well be quite different, especially at regional scales, from equilibrium

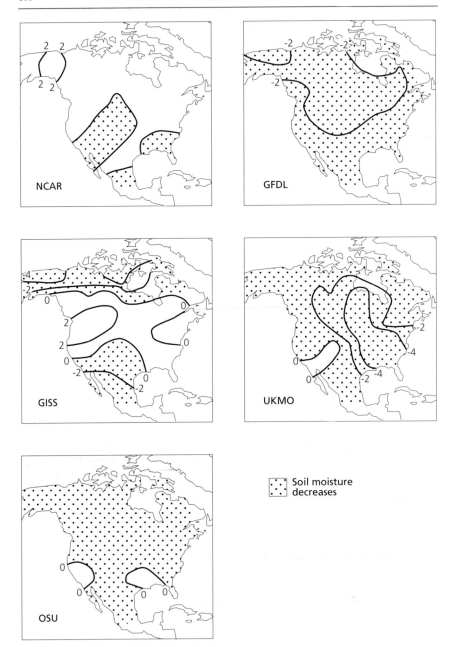

Fig. 4.5 July soil moisture for a doubling of CO_2 minus the control simulation for five different GCMs (in centimetres of soil moisture). Stippled areas show soil moisture decreased. Note the range of variation between the models. (After Schneider, 1990)

climates associated with particular greenhouse gas concentrations. So equilibrium GCM scenarios may be a poor guide to the climatic changes which could occur in the early phases of global warming. If the warming of about 0.5°C which has occurred since the mid-nineteenth century fitted in with the pattern of global warming produced by GCMs, one might expect that temperatures in higher latitudes would have risen substantially more than the global average. In fact there has only been slightly greater warming in these regions than in tropical areas. This could indicate that the increase of carbon dioxide and other greenhouse gases is not the cause of the warming trend, that the GCMs are wrong, or that the transient climate is different from the equilibrium one.

TIME-DEPENDENT MODELS

These problems have been tackled more recently by the use of **time-dependent models**, where increases in carbon dioxide concentration are phased-in gradually. Time-dependent scenarios can also build in the time lags caused by the effects of the oceans. The response of climate to transient forcing by greenhouse gases will be delayed due to the thermal inertia of the oceans, so in order to model transient climate, ocean heat storage and transport must be taken into account. Linking the rapid response time of the atmosphere to the very slow response time of the oceans is difficult, and validating coupled models is made harder because of the lack of observations over the oceans. In order to save computer power and time, ordinary GCMs treat interactions between the atmosphere and the oceans in a very simplified manner. More sophisticated time-dependent models link up or couple an atmospheric GCM with a dynamic model of ocean circulation, gradually increasing the levels of carbon dioxide in the atmosphere. They have produced some interesting differences in regional climates from the simple equilibrium GCMs in certain areas, especially the Antarctic and the North Atlantic, where levels of warming are reduced. But for many areas, such as Europe, the similarities are greater than the differences, though contrasts in precipitation patterns between the models are considerable (Rahmstorf, 1994).

Only a few coupled atmosphere-ocean models have been run so far due to the tremendous amount of computer power they require. For example, to run a ten-year phase of simulated climate may require one hundred hours of computer time, and this would have to be repeated many times for a single simulation. Because of constraints on computer power the models which have been run so far, rather than beginning from the climate and greenhouse gas levels of the pre-industrial period, have tended to start later – in some cases in the 1980s – assuming that the climate at this time was in equilibrium (Hulme et al., 1994). Few of these models have been run over time-scales of a century or two. In particular, more realistic modelling of the behaviour of the oceans requires a great deal of computer power due to the slowness with which the oceans respond to changes in the atmosphere. Coupled atmosphere-ocean models still do not build in important climatic feedbacks with the biosphere including variation in forest cover or the response of peat bogs in high northern latitudes to warmer conditions.

Recent time-dependent models, incorporating realistic forecasts for changes in greenhouse gas emissions from various sources, have predicted that a temperature rise of +1.1°C±0.5°C above the pre-industrial level should have occurred by 1990 (rather more than has actually been recorded) or 2.6°C±1.1 by AD2050. This involves an average warming to the middle of the next century of 0.26°C per decade, a much faster rate of change than during the last century.

The broad features of global warming produced by equilibrium GCMs are also identified in time-dependent coupled atmosphere-ocean models but, nevertheless, confidence in their portrayal of regional climates is still fairly low. The overall rates of warming, about 0.3°C per decade, are in line with earlier models. Transient change atmosphere-ocean models have sometimes produced interesting and potentially worrying differences from equilibrium GCMs. One would expect that, with a transient model, the global climate at the time when effective carbon dioxide doubling was reached would be less different from that of the present day than the equilibium climate for the same greenhouse gas increase which would ultimately occur, but that the changes would be in the same general direction and with similar large-scale patterns. However, recent experiments with coupled atmosphere-ocean models have not shown the enhanced warming in higher latitudes which equilibrium GCMs display. Does this mean that the results of the equilibrium GCMs – on which so much discussion of global warming is based – are misleading and wrong? Or is it that regional climates in the early stages of response to global warming may differ markedly from later ones? Better models coupling time-dependent GCMs to more realistic models of the oceans are being developed and may, in time, produce more accurate scenarios of regional climates. However, some models have already been developed, building in cumulative increases of various greenhouse gases at ten-year intervals. This allows the changing contribution of different greenhouse gases to radiative forcing to be estimated.

Model experiments have suggested that, with a ×2 carbon dioxide increase, there should be a decrease in cloud cover at low levels and an increase of high-level clouds such as cirrus which would produce a general warming effect. There are indications, however, that the scale of positive feedback caused by changes in clouds would be greater at low latitudes than in high ones, in the northern hemisphere at least. This might greatly reduce the latitudinal gradient in temperature increase predicted by most

GCMs with high latitudes warming more than low ones. This, in turn, might affect the pattern of atmospheric circulation and hence cloud cover, illustrating the complexities involved.

GREENHOUSE GAS EMISSION SCENARIOS

The use of time-dependent atmosphere-ocean GCMs involves the assumption that we can predict future levels of greenhouse gas emissions with some accuracy (Fig. 4.6). It is worth pausing for a moment to consider how true this is. Prediction of emissions scenarios for various greenhouse gases involves assumptions regarding future population growth, economic development, political changes and technological advances. In the 1990 IPCC report, four scenarios were produced for future anthropogenic greenhouse gas emissions (Houghton et al., 1990). The first of these assumed that no significant measures would be taken to try to reduce emissions. This was known as the **Business as Usual** (BaU) scenario. The other three involved an increasing degree of control on the growth of emissions.

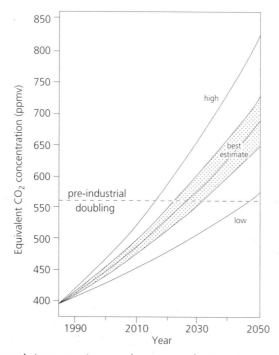

Fig. 4.6 Projected increases in greenhouse gas forcing (in equivalent carbon dioxide concentration). (After Warrick and Farmer, 1990)

The BaU scenario assumed that energy production from fossil fuels continued to use a substantial proportion of coal but that increases achieved by energy-saving technology were modest. Destruction of tropical rain forests continued until they were virtually eliminated, and no efforts were made to reduce emissions of methane and nitrous oxide. For the CFCs it was assumed that the Montreal Protocol was followed, but not by all countries. The predicted rises in atmospheric concentrations of the various greenhouse gases is shown in Table 4.1. Under the BaU scenario, carbon dioxide and methane levels reach more than double their present concentrations by the end of the next century, while nitrous oxide and CFCs continue to increase substantially.

IPCC Scenario B involved a shift in energy supply towards fuels containing less carbon, especially natural gas, with larger increases in energy efficiency. Deforestation was halted and the Montreal Protocol was fully adhered to. In this scenario the reduction in carbon dioxide and methane is especially impressive with less change in levels of nitrous oxide and CFCs (Table 4.2).

In Scenario C there is a shift towards renewable energy and nuclear power in the second half of the next century. CFCs are phased out and emissions of greenhouse gases from agriculture halted. Deeper cuts in carbon dioxide emissions slow the rate of growth of its atmospheric concentration, while levels of methane and CFCs start to fall towards the end of the century (Table 4.3).

Scenario D incorporates a move to renewable energy and nuclear power in the first rather than the second half of the next century. It indicates that tight curbs on emissions in the developed countries, along with moderate growth of emissions in developing countries, could stabilize carbon dioxide levels by the middle of the century along with substantial falls in methane and CFCs (Table 4.4).

The four emissions scenarios span a wide range of possible political responses, the feasibility of which will be considered in Chapter 7. In moving from levels of emissions of greenhouse gases to their atmospheric concentrations, the assumption is made that we know enough about their various sources and sinks to be able to relate the two closely. As noted in Chapter 3, however, the sources and sinks of some greenhouse gases are not yet fully known. These scenarios are confined to anthropogenic greenhouse gas emissions and do not incorporate the possibility that there may be substantial indirect effects through changes in natural sources and sinks, such as the release of methane from high-latitude wetlands (Chapter 3).

Moving from estimated future emissions of greenhouse gases to atmospheric concentrations and then to future global mean temperatures is a further uncertain step. The best estimate temperature rise for the four IPCC emissions scenarios are shown in Table 4.5. The potential sources of error in such calculations are obvious but Table 4.5 demonstrates that,

Table 4.1 Changes in greenhouse gas concentrations under the IPCC business as usual scenario

Gas	Current emissions	AD2030	AD2050	AD2100
CO$_2$	355 ppmv	460	530	825
CH$_4$	1720 ppbv	2750	3250	4000
N$_2$O	310 ppbv	345	370	415
CFC11	300 pptv	500	550	625
CFC12	550 pptv	900	1100	1400

(After Houghton et al., 1990)

Table 4.2 Changes in greenhouse gas concentrations under IPCC scenario B

Gas	Current emissions	AD2030	AD2050	AD2100
CO$_2$	355 ppmv	400	450	560
CH$_4$	1720 ppbv	2250	2500	2800
N$_2$O	310 ppbv	335	345	375
CFC11	300 pptv	480	500	550
CFC12	550 pptv	890	1000	1270

(After Houghton et al., 1990)

Table 4.3 Changes in greenhouse gas concentrations under IPCC scenario C

Gas	Current emissions	AD2030	AD2050	AD2100
CO$_2$	355 ppmv	400	450	500
CH$_4$	1720 ppbv	2000	2100	1600
N$_2$O	310 ppbv	315	340	375
CFC11	300 pptv	275	200	100
CFC12	550 pptv	600	550	350

(After Houghton et al., 1990)

Table 4.4 Changes in greenhouse gas concentrations under IPCC scenario D

Gas	Current emissions	AD2030	AD2050	AD2100
CO$_2$	355 ppmv	390	400	440
CH$_4$	1720 ppbv	1750	1700	1520
N$_2$O	310 ppbv	325	340	375
CFC11	300 pptv	275	200	100
CFC12	550 pptv	600	550	350

(After Houghton et al., 1990)

Table 4.5 Estimated increases in global mean
temperature for the four IPCC scenarios

Scenario	AD2030	AD2500	AD2100
BaU	+1°C	+1.75°C	+3.3°C
B	+0.6°C	+1.0°C	+2.0°C
C	+0.5°C	+0.9°C	+1.4°C
D	+0.5°C	+0.7°C	+1.0°C

(After Houghton et al., 1990)

even with marked emissions reductions (Scenarios C and D), on the basis
of present knowledge, global mean temperatures still seem set to rise by
at least 1°C, a modest figure but still twice the rate of the last century.
The 1992 IPCC update included a number of modifications to its
emissions scenarios based on additional knowledge of greenhouse gas
sources, such as revised rates of tropical deforestation, and political developments such as changes in eastern Europe and the collapse of the USSR,
as well as incorporating new government initiatives which were aimed at
reducing emissions, including the London amendments to the Montreal
Protocol. The range of possibilities which these scenarios demonstrated
was even wider than those in the 1990 report. The aim was to produce
a range of plausible assumptions which would help policy-makers weigh
up the potential advantages of implementing various strategies regarding
greenhouse gas emissions, and the scale of change which might be needed
to bring about given levels of emissions reduction. The feasibility of adopting various sets of policies and their economic cost is a topic which will
be considered in Chapter 7.

GCM SCENARIOS FOR FUTURE CLIMATE

Many GCM experiments, though differing in detail, have common
features regarding future climatic scenarios. They all show a significant
increase in average global temperatures with an equivalent doubling of
carbon dioxide. While the range of increase has been between 1.9 and
5.2°C, most have been between 3.5 and 4.5°C. The most recent models,
with more detailed resolution, have produced estimated temperature rises
towards the lower end of this range. It is widely believed by scientists that
the response of the climate system to an eventual doubling of equivalent
carbon dioxide levels is likely to be in this area. Rates of warming during
the next century have been estimated at around 0.3°C per decade
(Houghton et al., 1990) though recent work has suggested a lower figure.

Even so this is a much greater rate of temperature increase than has occurred during the postglacial period, emphasising that it is not so much the scale of climatic change which poses a potential threat but its speed, which may make adaptation difficult (Chapter 6). There is a general agreement in the models that future warming will be greater in higher latitudes, possibly 50–100 per cent above the global mean, and greater in winter than in summer. High-latitude warming in winter is predicted to reduce the extent of arctic sea ice and the area covered by snow, as well as encouraging faster spring snowmelt. On the other hand, the presence of summer sea ice and ice caps is predicted to reduce increases in summer warming to less than the global average in polar areas.

Warming in the tropics is expected to be less than the global average. In contrast, most models show above-average warming over the mid-latitude continental interiors of the northern hemisphere during summer. Precipitation is generally predicted to increase in mid- and high-latitude continental interiors during winter – perhaps by between 5 and 10 per cent – and in the tropics throughout the year. The monsoon circulation should move polewards. While GCMs are considered to provide a reasonable broad-brush picture of the likely range of possible future climates, particularly as regards temperatures, they do not indicate whether there is likely to be any change in the variability of temperatures.

At a global scale, GCMs suggest a net increase in precipitation of perhaps 7–15 per cent but regional variations are much less certain, possibly ranging from –20 per cent to +20 per cent. GCMs show a general increase in soil moisture in winter in most mid- and high-latitude areas but a decrease in summer over much of North America and Asia. The increase in summer dryness is especially great over the American Midwest and Great Plains. The increased soil moisture deficit in summer was caused by the earlier disappearance of winter snow cover, leading to an earlier start to evaporation from the soil, and by changes in mid-latitude precipitation patterns due to a poleward shift of rain belts.

As well as the uncertainties relating to future greenhouse gas emissions and the problems involved in trying to simplify the complex sets of interactions which exist within the climate system, there are other difficulties due to our lack of knowledge of the atmosphere. These incude how climate will actually respond to increases in greenhouse gases or the climate's sensitivity, and the various sinks for carbon dioxide, methane and other greenhouse gases (Chapter 3). This problem is compounded by the likelihood that the efficiency of the operation of these sinks, as well as the natural sources for gases like carbon dioxide and methane, will themselves be influenced by climatic change. In particular there is uncertainty about clouds and the ways in which they influence the radiation balance and, concerning the oceans, especially how they exchange energy with the atmosphere, and how they circulate it within themselves, horizontally and vertically. These influences may interact to modify current projections substantially. The consensus view at the moment is that the net result of such modifications is likely to increase rather than reduce temperature rises.

Although GCMs show reasonably close agreement on the scale of global temperature changes which may occur in the future and, to a lesser extent, changes in precipitation, there is far less agreement between them concerning climatic change scenarios at regional scales. Yet it is at precisely this level that we need to be able to make estimates of likely climatic changes because the effects of global warming on mankind will be articulated through variations in regional climates. However, as models are improved and the quality of their resolution increases, there is less disagreement and increasing consensus between them. Schneider (1990) has suggested that some of the likely regional climatic changes resulting from global warming may be warmer conditions and longer growing seasons in high latitudes, wetter subtropical monsoons, wetter winters and springs in high and mid-latitudes, drier summers in some mid-latitude areas, and more intense tropical cyclones. Some of the likely impacts resulting from regional scenarios of climatic change are examined in Chapter 7. To demonstrate some of the regional variations which may result from global warming and also to highlight the amount of unreliability which currently exists, a more detailed regional study of possible future climate in Britain is now considered.

Warrick and Barrow (1991) have produced some future climate scenarios for the UK, based on combining the average regional results of five equilibrium GCMs with a simple global time-dependent model and the IPCC BaU scenario for future greenhouse gas emissions. For summer, the average temperature change for the UK was similar to that for the global mean. Summer temperature rises were relatively uniform throughout Britain with an increase of around 0.7°C by 2010, 1.4°C by 2030 and 2.1°C by 2050. Because the temperature increase was uniform, the current level of difference in summer between southern England and northern Scotland would remain.

For winter, a much more marked temperature gradient was evident, reflecting the tendency of GCMs to show increased warming in higher latitudes. Under this scenario, northern Scotland would warm by about 1.4°C by 2010, 2.8°C by 2030 and 4.2°C by 2050 as opposed to 0.7°C, 1.4°C and 2.1°C respectively for southern England. The higher warming in northern areas would tend to even out current differences in winter temperatures. By 2030 central England would have winter temperatures comparable with present-day Bordeaux.

Assuming that the variability of climatic conditions remained unchanged, cool summers would become less frequent. One like that of 1972, which currently has a probability of occurring about eight times a century, would by 2030 only occur once in every 500 years. Warm summers like 1988–9 which only occur about twice a century now, would occur once every five years by 2030. Very cold winters like 1962–3 would have a negligible chance of occurring at all.

Scenarios for precipitation changes in Britain are far less convincing. The GCMs used in Warrick and Barrow's study had problems in simulating present-day precipitation and showed wide differences in their estimates for the future. All the GCMs agreed, however, that precipitation in the UK in autumn, winter and spring would increase, though there was no agreement by how much. For the summer season, critical as far as agriculture is concerned, three of the models suggested a decrease in precipitation and two of them an increase. Overall the scenarios for summer precipitation change were 0±5 per cent for 2010, 0±11 per cent by 2030 and 0±16 per cent by 2050. There was also a suggestion that there might be a greater chance of drier summers in future in the south of Britain compared with the north.

But, even if precipitation did not change, higher temperatures would tend to lead to a reduction in soil moisture due to increased evaporation. To take a worst-case scenario, if by 2050 summer precipitation had fallen, on average, by around 16 per cent, the chance of two very dry years together, like 1975–6, would be increased twelve times. Under such a scenario, recent droughts in the UK, like that of 1988–92, might become a more permanent problem for farmers and water companies (Marsh and Monkhouse, 1993). On the other hand, if summer precipitation increased by 16 per cent the chances of two such dry years occurring together would become negligible. For winter, the averaged results of the GCMs suggested an increase in precipitation of around 5 per cent by 2030 and around 15 per cent by 2050, fairly evenly distributed throughout Britain.

It is often claimed that, if global warming occurs as predicted, temperatures in Britain may rise to approach those of present-day Italy or Spain and that, as a result, the UK would enjoy a Mediterranean climate. Warrick and Barrow make the important point that this analogy is misleading. Many of the basic factors which control the British climate, such as its latitude, will remain unchanged and it is unlikely that the subtropical high pressure cell which brings dry conditions to the Mediterranean in summer would reach sufficiently far northwards to dominate Britain. Whatever future climate the British Isles may experience as a result of global warming it is likely to be unique, and without an exact contemporary analogue.

CLIMATIC VARIABILITY

GCMs suggest that a warming of 0.3°C per decade may be likely in the future but, for many people, rises in average temperatures of this magnitude are difficult to appreciate and are much less significant than possible changes in the incidence of extreme weather events. GCMs are poor at providing information on potential changes in the frequency of extreme events, especially relating to precipitation (Smith, 1993). There is no reason to believe that global warming will necessarily increase the range of climatic variability; in some areas, day-to-day variations may actually be less than at present. A wide range of unusual and extreme weather events in recent years has been linked by the media to man-induced global warming but these links have not been proved to the satisfaction of most climatologists. The IPCC report suggested that in mid-latitudes there might be a reduction in variations in the amount of windiness with low pressure systems becoming less intense. But even if the distribution of extreme events around the mean remains unaltered, many regions in a warmer world would experience an increased incidence of heatwaves while, by the same token, the frequency of cold spells would drop markedly.

Small upward shifts in average temperatures could have major effects in altering the frequency of hot conditions. For China, GCMs suggest that temperatures by AD2050 may be higher than extremely warm years during the last decade (Hulme et al., 1994). A slight shift in mean surface temperature could increase the probability of a bad drought from one year in twenty to one year in four – or a more severe one from one year in one hundred to one year in ten. Events which in the past would have been unlikely to occur once in a lifetime might become commonplace. On the other hand, if global warming is linked to greater temperature increases in higher latitudes than in tropical areas, reducing the temperature gradient between the equator and the poles, this may affect the speed of the westerly circulation in the northern hemisphere and possibly the direction of storm tracks, leading to the more frequent occurrence of blocking patterns caused by persistent high pressure systems. This could lead to more severe droughts in some mid-latitude areas including Europe.

For the tropics, however, there is the worry that, with higher average temperatures, cyclones may become more frequent and more severe. The geographical occurrence of these storms is related to the occurrence of sea surface temperatures of 26–27°C and above. If the oceans became warmer and these conditions more widespread, then areas which have not experienced tropical cyclones in the past may become vulnerable to them, while warmer conditions might also extend the length of the hurricane season in areas which already experience cyclones.

DETECTING THE SIGNAL THROUGH THE NOISE

If global warming proceeds at something approximating to the rate suggested by GCMs, when will it be possible to say that it is definitely occurring? This question is extremely hard to answer. It is one thing to say that particular climatic patterns and trends are in line with those predicted to result from the man-enhanced greenhouse effect. It is quite another to prove that they are the direct result of greenhouse gas forcing. To understand why, it is worth recalling that, in the northern hemisphere, seasonal changes in average temperature between winter and summer are around 15°C. This is some thirty times the increase in average global temperatures that is thought to have occurred during the last century. The likely long-term changes in mean annual temperature due to global warming, even taken over several decades, are still considerably smaller than the range of short-term variation. Short-term natural variability in the climate system is so great that it makes it difficult to identify a gradual rising trend due to enhanced greenhouse gas forcing. The amount of background 'noise' is so great that it hides any clear global warming 'signal'. If the twentieth-century warming which has been observed has been due entirely to the man-enhanced greenhouse effect then the fact that the temperature rises which have occurred are lower than predicted may indicate that the sensitivity of the climate to changes in greenhouse gas concentrations is smaller than has sometimes been believed. But it is also possible that natural variability within the climate system may have offset or damped down temperature rises caused by the enhanced greenhouse effect.

Efforts have been made to 'massage' existing temperature data by removing or 'factoring out' variations due to identifiable causes, such as volcanic eruptions and ENSO events. However, focusing on global mean temperature as the sole indicator of possible man-induced global warming is a rather crude approach. A more complex exercise involves looking for what have been called 'fingerprints' left in the temperature record by climatic changes, due to rising greenhouse gas concentrations. The fingerprint is a signal whose structure is unique to the expected man-enhanced greenhouse effect. For instance, if GCMs suggest that overall temperature increases should be accompanied by greater warming in higher latitudes than the tropics, greater warming in the northern hemisphere than the southern due to the greater extent of land in the former and sea in the latter, or greater warming over the continents than over the oceans, we can examine climatic records in more detail to see if such trends exist and can be isolated.

A more detailed look at temperature trends in the Arctic shows how easy it is to oversimplify concepts such as high-latitude warming (Walsh, 1993). There is no evidence of significant warming in recent decades in the high arctic, on the basis of drifting ice stations mainly above 80° north. In fact there has been a significant decrease in surface temperatures in the western Arctic. However, when land-based stations at less high latitudes in the Arctic are studied, a different pattern emerges. There is evidence of warming in the last thirty years from Alaska, Northern Canada and

Siberia as well as over the ocean north of Spitzbergen. Warming in some
of these areas has been in excess of 0.5°C per decade. In contrast, there
has been a marked cooling of between 0.25–0.5°C per decade in an area
south of Greenland. Strikingly similar patterns of temperature change have
emerged from recent time-dependent coupled atmosphere-ocean models.

The problem with this approach is that the regional scenarios produced
by GCMs are not yet sufficiently detailed to give a clear image of what
this 'fingerprint' should look like, or to identify areas of the world whose
climates might be especially sensitive to changes associated with greenhouse gas forcing. In addition, we do not have long enough runs of
suitable data to do this. Data on surface air temperature and precipitation have been collected for several decades in most parts of the world
but for other key variables, such as concentrations of sulphate aerosols in
the atmosphere, or arctic sea ice limits, data only go back a few years.
Most of our information is collected at the Earth's surface; data from
satellite observations are only available for a limited period. The problem
of identifying a signal is even worse for precipitation than temperature
because projections of future changes by GCMs are thought to be much
less reliable.

Recent research on the pattern of vertical temperature changes in the
atmosphere since the 1960s shows that they correspond closely with those
produced by the equivalent doubling of carbon dioxide scenarios in
GCMs, with a tendency for warming in the upper troposphere and cooling
in the stratosphere. But other climate forcing mechanisms, such as ozone
depletion, may be contributing to stratospheric cooling as well as changes
in greenhouse gases. The impact of atmospheric aerosols on global climate
is being seen as increasingly important. The complex influences of aerosols
will make detection of a 'greenhouse fingerprint' much harder and will
need to be incorporated into future GCMs (Taylor and Penner, 1994 and
Wigley, 1994). Longer runs of data will be required before we can be
more certain. Optimistic estimates suggest that at least another decade or
two of increasingly sophisticated research will be needed before we can
prove that global warming is, or is not, happening. Some climatologists
suggest that thirty to forty years is a more realistic forecast.

5

The Impact of Global Warming: Sea Level Rise

Before we look at the possible impacts of global warming it is worth emphasising that other human activities and environmental changes will have an equally significant or even greater effect on mankind in many parts of the world in the next century or so. The impacts of global warming are, however, likely to reinforce this. Of all the projected impacts of global warming, the threat posed by rising sea levels is probably the easiest to appreciate. This does not mean, however, that it is easy to predict what will happen to sea levels on a time-scale of decades if global warming occurs roughly as predicted. There has been more scaremongering and alarmist reporting in the media concerning the prospect of rising sea levels than any other aspect of global warming because redrawing coastlines on the basis of a higher datum for mean sea level is a deceptively simple exercise. It may be difficult to imagine the possible impact of a 3°C rise in mean global temperatures but it seems easy to visualise the effects of a 3 m rise in sea level. The environmental lobby has produced some extreme projections of sea level rises of up to 5–6 m. These have been presented – or, at least, have been interpreted by many people – as certainties rather than merely possibilities which are most unlikely to occur within the next few decades. Rising sea levels have become the problem most popularly identified with global warming.

MEASURING SEA LEVEL RISE

Climatic warming can affect sea levels in three main ways. Differences in wind patterns may change the frequency and severity of high water levels caused by storm surges. Warmer conditions will cause a rise in mean sea level due to the thermal expansion of sea water. Finally, higher temperatures will cause a reduction in the volume of the world's glaciers and ice caps leading to a further rise in sea level.

It is easy to forget that, like temperature changes, sea level variations are nothing new. During the last interglacial, some 120,000 years ago, sea levels were about 5–6 m higher than they are now, possibly due to the melting of Antarctic ice. At the maximum of the last glaciation they were about 140 m lower due to large quantities of water being locked up in ice sheets. At the end of the last glaciation, with the return to the sea of vast quantities of meltwater, sea levels rose. Following the break up of the Laurentide ice sheet around 8,000 years ago, sea levels rose by more than 46 mm per year for short periods (Tooley and Jelgersma, 1992). Mean sea level continued to rise by 6–12 mm per year up to 6,000 years ago. Since then it has carried on rising, but more slowly, corresponding with the stability of the Holocene climate; by 1.0–1.4 mm per year and, within the last 1,000 years, by around 0.1–0.2 mm per year. Theoretically, with the increase in greenhouse gas concentrations which has occurred in the last century, and the observed rise in global surface temperatures, there should have been an acceleration in the rate of sea level rise. Whether or not this has actually occurred is unclear. Certainly it has not yet been detected from tide gauge records (Woodroffe, 1994). It may take another three or four decades before this becomes possible, though satellite monitoring of sea level changes may provide an answer more quickly. Projected sea level rises in the next few decades have been placed at 1.0–2.0 mm per year, an order of magnitude greater than those experienced during historical times.

So far this book has discussed synchronous worldwide or **eustatic** changes in mean sea level. The picture of sea level change relative to the land has been very different from this in certain areas. In north-west Britain and Scandinavia, while sea levels rose after the last ice age, the land, relieved of the weight of the ice, rebounded even further. Around the coasts of Scotland, sets of raised beaches bear witness to the scale of isostatic recovery. Although they appear to be horizontal relative to present sea level, careful measurement shows that they rise steadily towards the areas where the ice sheet was thickest and postglacial uplift has consequently been greatest. Isostatic recovery continues today though at a reduced rate compared with early postglacial times. At the head of the Gulf of Bothnia the land is rising by around 1 cm each year, and Finland acquires as much as 1,000 km^2 of new land each century. Parts of western Scotland are still rising by up to 6 mm each year (Boorman et al., 1989).

However, in areas just beyond the maximum extent of the ice sheet margins, the land was pushed up into a forebulge. The gradual collapse of this since the end of the last glaciation has produced subsidence in south-east England, The Netherlands and parts of eastern North America. Sea level at New York has risen by twice the global average during the last century.

In other parts of the world, tectonic activity has led to slow land subsidence and major delta areas are also sinking under the weight of deposited sediments. At a local scale, relative sea level is sometimes changing due to human activities. Parts of Hong Kong are sinking as the weight of highrise buildings compacts the marine sediments on which they rest (Yim, 1993). Bangkok is subsiding some 10–15 cm a year due to the pumping of groundwater (Bird, 1993). The problems of Venice, built on wooden piles set into sandbanks, are well known. About half the relative sea level rise of around 30 cm recorded at Venice during the last century is due to man-induced subsidence caused by groundwater removal. Venice subsided by about 10 cm between 1952 and 1969 as a result of this. The rate of sinking has been reduced since pumping was stopped but, when man-made subsidence is added to natural sinking and the effects of eustatic sea level rises, Venice, only 0.8 m above mean sea level, is extremely vulnerable (Fig. 5.1). Figure 5.2 shows those areas of the world's coastline which have been subsiding in recent years and which are particularly vulnerable to rises in mean sea level resulting from global warming. Agriculture can have an impact as well. Parts of the English Fens have been lowered from 1.6 m above Ordnance Datum in 1848 to 2.3 m below it in 1979 due to the drying and shrinking of the peat surface (Tooley and Jelgersma, 1992).

On shorter time-scales, variations in atmospheric pressure and wind patterns can affect sea levels. Sea level can rise 1 cm for every millibar fall

Fig. 5.1 High tide, Piazza San Marco, Venice demonstrating the city's vulnerability to even a small rise in mean sea level. (Photo: I. Whyte)

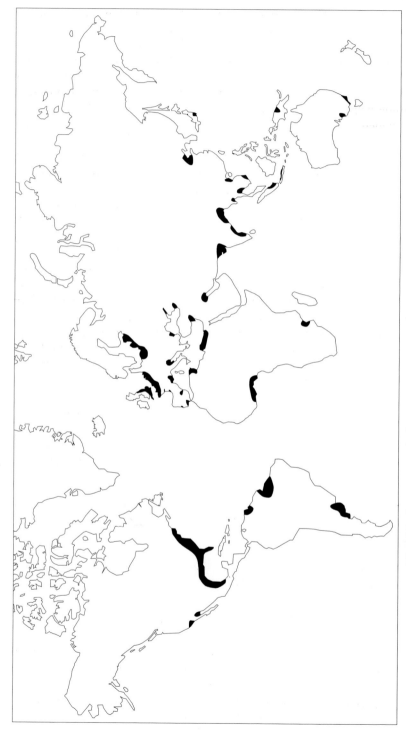

Fig. 5.2 Areas particularly vulnerable to sea level rise. (After Bird, 1993)

in pressure. Storm surges associated with Atlantic depressions or tropical cyclones can produce marked variations in sea levels as do the tides over even shorter periods. Tidal ranges of up to 20 m have been recorded. On longer time-scales sea level is influenced by ENSO events. The 1982–3 ENSO event raised sea level 20–30 cm along the Pacific coast of North America. The existence of so many short-term variations of considerable amplitude helps explain the difficulty of identifying long-term changes.

It is widely believed that global average sea levels have risen in the last century in step with temperature increases (Fig. 5.3). Estimates of the amount of rise have varied from 5–30 cm (Tooley and Jelgersma, 1992) but several suggest an increase of between 10 and 15 cm or 1.0–1.5 mm per year (Warrick and Farmer, 1990). However, as the various calculations have all been produced by manipulating the same basic data set of tide gauge records, their similarity should be treated with an element of caution.

About half the recent sea level rise is considered to have been due to thermal expansion, and half to the melting of glaciers and small ice caps which, although only containing about 0.3 per cent of the world's land ice, respond significantly to even minor changes in temperature. The Greenland ice cap is much greater in extent and there is uncertainty as to whether its mass balance has changed significantly in the last century, though some scientists believe that recent climatic warming has produced a negative mass balance with a net loss by melting and calving. If

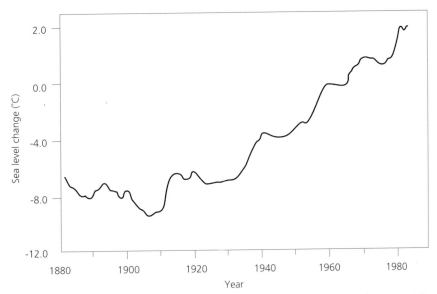

Fig. 5.3 Global mean sea level rise 1880-1990 relative to the average for 1951–70. (After Roberts, 1994)

Greenland has contributed anything to recent sea level rise, that input has probably been small. It is uncertain how precipitation patterns over Greenland are likely to change with global warming and how this will affect the ice cap's mass balance. The Antarctic ice cap, for reasons which will be considered in a moment, has probably made no positive contribution to recent sea level rise.

Mean global sea level is expected to rise in the future with global warming. First, a 3°C rise of average global surface temperatures would increase sea level between 8 and 14 cm by thermal expansion alone. Second, global sea level can be expected to rise with the melting of smaller glaciers and ice caps (Kuhn, 1993). Sea ice, displacing its own volume of water, makes no contribution to sea level rise when it melts. Third, there is the possibility that the vastly larger Greenland and Antarctic ice caps might start to melt under warmer conditions.

Estimates of future sea level rise have a considerable range of uncertainty due to the problems of estimating future greenhouse gas concentrations and changes in global mean temperatures which were considered in Chapter 4, plus the difficulty of evaluating the impact of temperature and precipitation changes on the mass balance of glaciers and ice caps, and the response of oceans to temperature changes (Fig. 5.4). Most recent

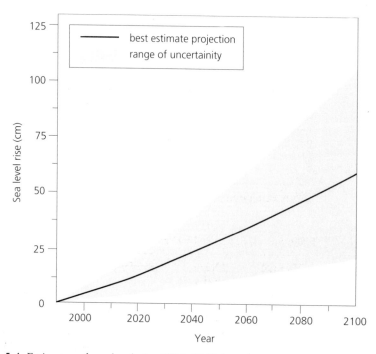

Fig. 5.4 Estimates of sea level rise 1990–2100 based on a modified version of the IPCC Business-as-Usual scenario. (After Mintzer, 1992)

studies have suggested a rise of between 10 and 30 cm by 2030, or about
58 cm by 2100, though with an uncertainty range of 21–105 cm. The
IPCC report suggested a possible rise of up to 65 cm (±30) over the next
century under its BaU scenario (Table 5.1). Warrick and Farmer's (1990)
study of possible sea level changes has suggested that, between now and
2030, sea level is likely to increase between 17 and 26 cm. The effect of
sea level rise due to the melting of Greenland ice would be approximately
balanced by the impact of Antarctica in reducing sea level (Table 5.2).

During the last interglacial, about 120,000 years ago when the world
was around 2°C warmer than it is at present, mean sea level seems to
have been about 5 m higher than now. On the other hand, if mean temperatures rise by 2°C by AD2050 this does not automatically mean that sea
levels will be 5 m above their present extent. First, a substantial time lag
is involved, especially with the response of the oceans and the Greenland
and Antarctic ice caps to warming, a time lag which could involve
hundreds, even thousands, of years.

Measurement of eustatic sea level change is more difficult than might be
imagined. Calculations of long-term changes in mean sea level depend on
the analysis of sets of tide gauge records (Gornitz, 1993 and Pugh, 1993).
Mean sea level is defined as the mean of the height of the sea surface
measured at hourly intervals over a long period of time – traditionally
nineteen years – but it can be considered more simply as the long-term

Table 5.1 Estimates of sea level rise associated
with IPCC scenarios

Scenario	AD2030	AD2050	AD2100
BaU	20 cm	30 cm	65 cm
B	18	20	44
C	17	18	40
D	16	18	32

(After Houghton et al., 1990)

Table 5.2 Possible sea level changes 1985–2030

	Thermal expansion	Alpine glaciers	Greenland	Antarctica	Total
Low estimate	4.3 cm	2.2	0.8	−2.3	5.0
Best estimate	8.4	8.3	1.8	−1.9	16.6
range	14.0	12.6	3.1	−3.3	26.4
High estimate	17.8	19.2	3.9	−3.0	43.9

(After Warrick and Farmer, 1990)

average of high and low tide levels. A long time series is required to remove the influence of tidal cycles – not just diurnal ones but the 14.6-day spring/neap tide cycle and the 18.6-year cycle based on the precession of the moon's orbit.

Tide gauges are far from uniformly scattered around the world's coasts. There is a marked concentration on Europe and North America and only a thin scatter around many coasts in the southern hemisphere. Individual sets of data from particular tide gauges must be scrutinised in order to pick out local changes due to tectonic uplift and subsidence or isostatic recovery. One technique is to reject sets of data from known areas of crustal instability. Another is to assume that, if sufficient data points are available, the positive and negative anomalies due to purely local circumstances will cancel out.

Most analyses of sea level rise are based on data collected by the Permanent Service for Mean Sea Level which computes information from nearly 1,400 tide gauges around the world (Fig. 5.5). However, many tide gauge records are only available for less than twenty years. Establishing long-term trends in sea level is difficult because a high percentage of tide gauge records go back for such short periods. A very high percentage of all gauges are in the northern hemisphere (though it could be argued that so are many of the world's coastlines). A wider network of recording stations, the Global Sea Level Observing System (GLOSS), is being developed which should, in the future, provide better data for monitoring global sea level changes (Warrick et al., 1993). As with changes in mean annual temperature, short-term fluctuations in sea level caused by tides, storm surges and larger-scale pressure variations are of far greater amplitude than long-term eustatic changes due to variations in mean global temperature.

Recently it has been suggested that some human activities unrelated to climate may have contributed to past sea level rise (Sahagian et al., 1994). The abstraction of groundwater from aquifers which do not recharge under present climatic conditions represents a net return of water to the oceans, as does the loss of moisture over land areas due to desertification, or losses from inland water bodies such as the Caspian and Aral Seas. The use of water from aquifers in the western part of the USA may be contributing 0.1 mm a year to sea level rise. The Aral Sea, affected by the diversion of water for irrigation, has dropped 16 m and lost half its surface area since 1960, perhaps contributing 2.2 mm to sea level rise. Desertification in the Sahel has produced an estimated 0.28 mm sea level rise since the 1960s. Deforestation may also have accounted for a 3.4 mm rise. An overall sea level rise of 17 cm due to these causes has been suggested. To be set against this is the construction of reservoirs which withhold water from the sea. These are thought to have reduced sea levels by 5.2 cm. The combined effects produce a rise of around 11.8 cm from causes not directly related to global warming. Such estimates may not be

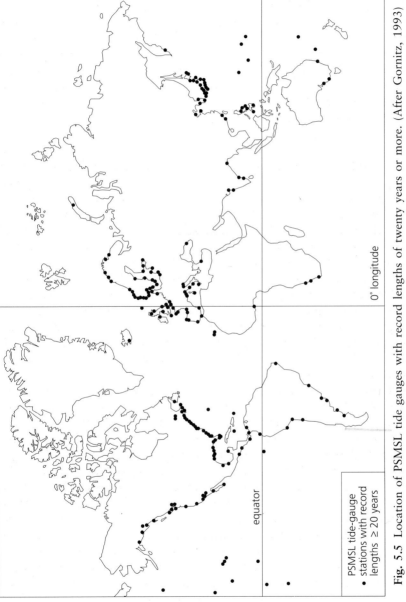

equator

0° longitude

PSMSL tide-gauge
• stations with record
lengths ≥ 20 years

Fig. 5.5 Location of PSMSL tide gauges with record lengths of twenty years or more. (After Gornitz, 1993)

accurate but they indicate the possibility that the contribution of thermal expansion and ice melting to sea level rise may have been overestimated.

Projections of the future behaviour of small glaciers and ice caps is complicated by the fact that, while their area is fairly easy to calculate, estimates of their volume vary considerably and the lack of detailed mass balance studies makes it hard to predict their response to warming. The Greenland and Antarctic ice caps between them contain 99 per cent of the world's ice. The Antarctic ice cap is ten times as large as the Greenland ice cap but they exist in quite different climates. The climate of Greenland is much warmer than that of the Antarctic. The Greenland ice sheet loses as much by melting as by the calving of icebergs. In the Antarctic, conditions are much colder and melting is insignificant. Losses are almost entirely by the calving of icebergs into the southern ocean. Because of the low temperatures the amount of precipitation and the rate of snow accumulation over most of the Antarctic is very low. Atmospheric warming in the Antarctic is likely to increase precipitation and snow accumulation but, in Greenland, higher temperatures are likely to lead to more rapid melting. Changes in the Antarctic peninsula, which pushes far to the north and is much warmer than the rest of the continent, may point to future trends for the Antarctic as a whole (Paren et al., 1993). Though only a small part of the Antarctic, the peninsula contains a volume of ice equal to all the world's small glaciers and ice caps combined. In the last forty years the area had experienced a 2°C rise in mean annual temperatures and a marked increase in the rate of snow accumulation.

Table 5.2 shows the estimated contributions to sea level rise between 1985 and 2030. Net accumulation on the Antarctic ice cap is projected to cause a decrease in sea level of 1.9–3.3 cm. The net effect of all these forces is a projected rise of sea level of between 17 and 26 cm, the effects of the Greenland and Antarctic ice caps roughly cancelling out, leaving 52 per cent of the net rise due to thermal expansion and 48 per cent to the melting of glaciers and small ice caps. A substantial increase in accumulation of snow would eventually work its way through the ice sheet to produce a correspondingly greater loss by calving, but so slow is the response of the Antarctic ice sheet that this would take thousands of years. The lowest levels of the Antarctic ice are probably only just beginning to respond to the temperature changes which occurred at the end of the last glaciation over 10,000 years ago.

THE IMPACTS OF SEA LEVEL RISE

In general terms, the impacts of rising sea levels are likely to include:

a Increased risk of flooding of coastal lowlands, especially reclaimed land
b Accelerated coastal erosion

c Increased risk of flooding from more frequent storm surges
d Problems of drainage of low-lying land
e Increased intrusion of salt water into groundwater, soils and freshwater ecosystems
f Threats to existing sea defences and port facilities
g Possible permanent loss of agricultural land
h Disruption of fisheries and coastal ecosystems
i Damage to or loss of recreational beaches
j Changes in river sedimentation.

Most impact studies have, however, been based merely on drawing a new set of coastal contours on the basis of a sea level rise of 1, 2 or 3 m. This is a simplistic approach, a first approximation of potential impacts which underestimates the complex and dynamic nature of coastal environments. In many cases changes due to a rise in mean sea level will be superimposed on changes which are already occurring in coastal systems due to other human activities whose impacts may be even greater. Some of these activities, such as pollution and the mining of coral in the Maldives (see below), may impair the ability of coastal systems to respond to and cope with changes in sea level. The study of the potential impacts of sea level rise have been based on various yardsticks. A 1 m rise within the next hundred years is a common one (Tooley and Jelgersma, 1992) but the effects of rises of 2–3 m have also been widely considered.

Prediction of the effects of sea level rise is made more difficult because many coastal areas have a long history of dense settlement and have already been heavily affected by human interference. There is a need to focus research on the specific problems likely to arise in particular areas. Sea level changes will have different effects depending on whether coasts are characterised by cliffs, beaches, estuaries, deltas, swamps, coral reefs or man-made structures. But even within each of these categories there is a wide range of variation, such as the susceptibility of cliffs to erosion or the nature of beach material.

Drawing revised contours to represent the impact of higher sea levels is only a rough exercise because coastal environments are dynamic systems which respond to change in complex ways. There is a need to develop models of coastal change which are time-dependent and give indications of how dynamic processes such as sedimentation rates, changes in ecosystems and land use might operate. Existing human interference could in many cases exacerbate the impact of future environmental changes. Ill-advised engineering works are accelerating the erosion of beaches, already made vulnerable by the sea level rises that have occurred within the last century, or encouraging cliff retreat. A better understanding of the processes operating in particular coastal environments can allow man to work with rather than against natural systems in responding to sea level rise and its effects.

On cliffed coasts, rising sea levels will cause more intensive wave attack at the base of the cliffs, leading to instability and more erosion in many areas. On rocky coasts where strata are unstable, sea level rise could trigger off major landslides though, once these had occurred in a particular locality, the fallen material might then help to protect the base of the cliffs from further attack. Parts of the east coast of England are retreating by about 200 m per century (Figs. 5.6 and 5.7). On the east coast of England, where some cliffs of till are already retreating on average 1 m a year, the retreat would, it is estimated, accelerate 0.35 m a year for every 1 mm increase in mean sea level (Clayton, 1993). In Norfolk, where parts of the coast have an offshore slope of 1:500, a 10 mm rise in sea level would push the coast inland by 5 m. On the other hand, increased erosion of the Holderness coast and transport of the eroded material southwards by longshore drift might bring in sufficient material to provide enough protection to stabilise the Norfolk coast.

Sea level rise will tend to widen and deepen estuaries, allowing tides to push farther inland. Higher sea levels will hold back rivers at periods of high discharge, producing more widespread floods. Salt water will penetrate further inland and seep into coastal fresh water aquifers, a trend which will be encouraged by the abstraction of groundwater for human use. Changes in precipitation will have indirect impacts on estuaries through their effects on river regimes. Increased run-off may threaten the stability of embankments while decreased streamflow could encourage salt water to push further upstream (Tooley and Jelgersma, 1992).

A sea level rise of only 30 cm, it has been calculated, would cause 15–30 m of erosion in the north east USA, 30–300 m in Florida and possibly up to several kilometres in the wetlands of Louisiana where recent sea level rises have already caused the break-up of offshore barrier islands, exposing the wetlands behind to erosion. Many beaches, such as those at Miami, are already being maintained only by the large-scale dumping of sand. The cost of replenishing major tourist beaches in the USA has been put at US $14 billion for a 50 cm sea level rise.

It is thought that 70 per cent of the world's beaches are already eroding; under 10 per cent look as if they are being built up and the rest seem stable (Bird, 1993). Whether beaches are eroding or accreting may be due to a range of local circumstances as well as eustatic sea level rise; for example a reduction in sediment supply due to man-made coastal changes. Sandy coastlines are likely to retreat 50–100 times the amount of sea level rise. As many beaches at coastal resorts are narrow then even a modest rise in sea levels, such as has been widely predicted to occur by AD2030, would totally remove many beaches unless they were replenished artificially.

Coastal wetlands in many parts of the world would be threatened by higher sea levels. They have developed during several thousand years of relatively stable sea levels and might not be able to respond to the rapid

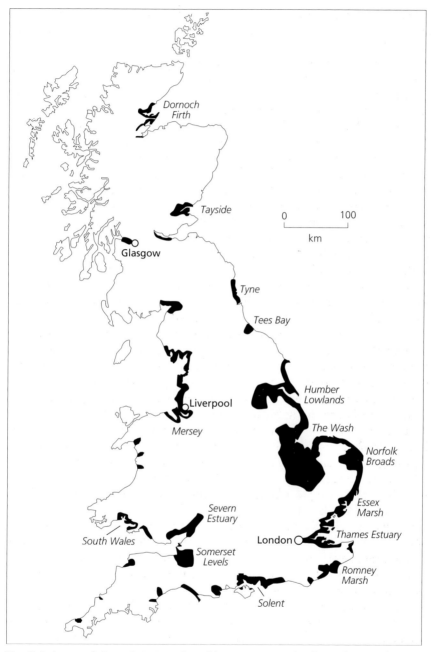

Fig. 5.6 Areas of Great Britain vulnerable to rising sea levels. (After Bird, 1993)

Fig. 5.7 Recent cliff erosion on the coast of Holderness. (Photo: Dr. A. Pringle)

changes which have been predicted. Left alone, salt marshes in areas such as south-east England can cope with rises in sea level of at least 5 mm a year but this involves a landward relocation of the marsh with erosion at the outer edge and accretion behind. When this process is blocked by coastal defences salt marshes may be in danger of disappearing entirely (Pethick, 1991). Salt marsh erosion has already been reported from many areas where sediment supply is insufficient to maintain accretion. Where the landward edges of salt marshes are defended by embankments, such marshes could be completely removed as sea levels rise. Salt marsh erosion in the Venice lagoon, due to subsidence and the effects of land reclamation, is already a serious problem.

In terms of the threat posed by flooding, The Netherlands might seem to be the most vulnerable country in Europe (de Ronde, 1993). Of the 14 million population of The Netherlands 8 million already live below mean sea level (Tooley and Jelgersma, 1992). In fact the Dutch dyke systems are already built to cope with substantially higher sea levels than the current average, for they have to contend with periodic storm surges. Following the floods of 1953, the heights of the new dykes were designed to cope with floods of a severity unlikely to occur less than once in 10,000 years. Even so a sea level rise of only 15 cm would increase the chances of storm surges exceeding the danger level by a factor of about times two. A more serious problem for the Dutch is likely to be increased penetration of saltwater up the Rhine. River water currently supplies about two thirds of the fresh water used in The Netherlands, so salt water intrusion is a serious problem. The costs of coping with sea level rise in The Netherlands have been estimated at around 20 billion guilders for raising dykes 2 m, 5–6 billion for checking beach and dune erosion, and 10 billion for water management costs due to rising sea level. Such an expense would, however, represent an annual cost of under 0.5 per cent of Dutch gross national product (Den Elzen and Rotmans, 1992).

The Mediterranean also faces distinctive problems relating to sea level rise (Fig. 5.8). Due to topography and a long history of over-exploitation and erosion, the coastal strip tends to hold a higher proportion of the population and its related activities than in northern Europe, a focus made ever more intensive by modern tourist developments. Even areas like the French coast between Sete and Perpignan which were shunned for centuries due to their malarial mosquitoes have now been developed as major tourist centres. In addition, the most productive agricultural areas are all in low-lying deltas vulnerable to flooding, while there is much less of a tradition of constructing coastal defences than in northern Europe. Many Mediterranean deltas are already suffering from erosion due to a reduction in river discharge and sediment supply resulting from water management schemes, afforestation and the diversion of water for irrigation. The building of a series of dams on the River Rhône to generate hydroelectricity has more than halved the river's sediment load. Erosion

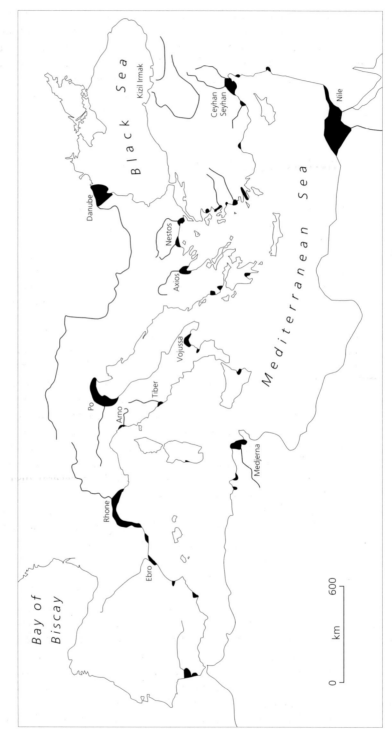

Fig. 5.8 Coastal lowland areas around the Mediterranean especially vulnerable to sea level rise. (After Jeftic et al., 1992)

rates of up to 40 m a year have been recorded recently in parts of the Rhône delta. In response to this, new systems of sea walls and other protective measures such as groynes have been constructed. Extrapolating from present trends, erosion by 2050, accelerated by rising sea levels, might be over 0.5 km in some areas unless suitable action is taken. Such erosion may weaken or remove existing dune belts and increase the risk of coastal flooding during storm surges (Jeftic et al., 1992).

IMPACTS ON DELTA AREAS

The world's great deltas are among the most densely populated and most vulnerable of coastal areas threatened by sea level rise (Broadus, 1993). In recent centuries many deltas have been growing due to deforestation and large-scale soil erosion in the upland parts of their catchments and intensive agriculture in the lowlands. In the Mediterranean, for example, this has caused substantial growth of the Ebro and Po deltas (Jeftic et al., 1992). Mining has also increased sediment yields in some South American and south-east Asian rivers.

Of all countries, Bangladesh is the most vulnerable to sea level rise, as nearly 80 per cent of its area consists of low-lying delta. Large areas of the Ganges-Brahmaputra delta in Bangladesh are less than 1 m above present mean sea level. It is important to appreciate that the main threat from rising sea levels in such areas is not the direct effects of sea levels creeping gradually upwards but the likelihood of more frequent, and possibly more severe, flooding episodes associated with storm surges on top of higher sea levels. Storm surges in the Bay of Bengal can rise 9 m above normal levels (Flather and Khander, 1993). During one storm surge in 1970 over 200,000 people died. There have already been several disastrous floods with massive losses of life in recent years, associated with storm surges caused by tropical cyclones; in 1985, 1987, 1988 and 1991 for example. The 1991 cyclone killed at least 125,000 people. Subsidence rates range from 10 mm a year to as much as 25 mm in some areas. In a country where population has been rising by 2.5 per cent per annum, parts of the delta support in excess of 1,000 people per km², with settlements occupying as much as 40 per cent of the land area (Brammer, 1993). A 1 m rise in mean sea level would flood 17 per cent of Bangladesh's arable land and threaten many additional areas with periodic floods.

Further sea level rise would be another problem to add to the difficulties of a dynamic and rapidly changing environment. In the recent past there have been major shifts in river courses, and dramatic changes in sediment loading. The 1950 earthquake in Assam released vast quantities of sediment which, when they had worked their way through the river system to the delta, raised river bed levels, increasing the flood risk, and created considerable areas of new land. Some of this new land is now being eroded, a trend likely to accelerate with rising sea levels, though in future this may be balanced by an increase in deposition due to erosion upstream in the Himalayas. There are also worries that flood protection schemes upstream could have major effects on sedimentation and fluvial processes which could worsen the effects of sea level rises.

Higher sea levels in the Bay of Bengal may also have serious effects on river hydrology; by altering the base level of rivers, they would encourage the deposition of sediment in the river channels, raising river channels above the surrounding areas and so increasing the risk of flooding. This risk might be greater inland than on the coast where accretion of new land might keep pace with sea level, emphasising the complex nature of the dynamic processes involved and the danger of taking too simplistic an approach. The increased risk of flooding might require the raising of villages, roads and railways, cut down the areas suitable for dry land crops, requiring a switch to different, lower yielding rice varieties, and increasing the penetration of salt water into groundwater and soils with further deleterious effects on crops, especially rice. Salt water intrusion already extends 150 km inland and this could increase with higher sea levels.

In Egypt erosion of the Nile delta has increased since the construction of the Aswan High Dam reduced the supply of silt by 90 per cent (Fig. 5.9). Since the construction of the dam, the river's discharge at the delta has dropped to only 4–6 per cent of its former level. Recent rates of erosion of the delta have exceeded 100 m a year in places. The Nile delta makes up only 2.3 per cent of the area of Egypt but has 46 per cent of the cultivated area and 45 per cent of the population. A quarter of the delta lies below 2 m with extensive areas between 1 m and 3 m below sea level and natural subsidence of up to 5 mm per year is already occurring. A 1 m rise in mean sea level could flood up to 20 per cent of the delta and displace 8–10 million people. A 1 m rise in sea level would flood up to 15 per cent of the nation's arable land and push the coastline inland by several kilometres.

The effects of sea level rise on delta areas will be influenced by the sediment loads of the rivers feeding them. As global warming proceeds some areas may experience greater precipitation, increasing the volume of sediment transported and deposited, while some areas may become drier and experience the opposite effect. Clearance of forest areas in the tropics is likely to increase the sediment loads of some rivers but, for some catchments in the Mediterranean, sediment yields are falling due to the fact that erosion in the past has stripped the higher parts of the catchments almost bare of soil.

The Mississippi delta already experiences a sea level rise of over 1 cm per year, most of it due to local subsidence. Before the twentieth century, the delta seems to have been in a state of dynamic balance with sediment

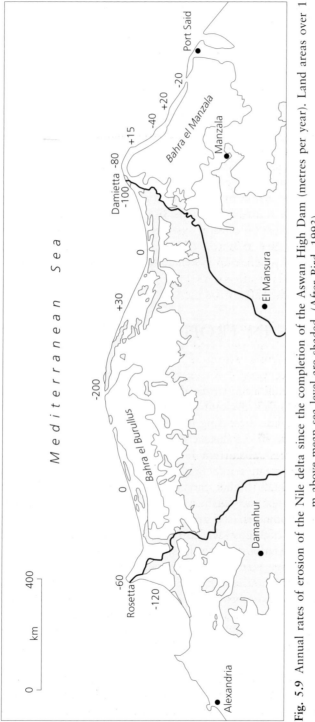

Fig. 5.9 Annual rates of erosion of the Nile delta since the completion of the Aswan High Dam (metres per year). Land areas over 1 m above mean sea level are shaded. (After Bird, 1993)

accumulation matching tectonic subsidence. Over the past 5,000 years the delta has been accreting by 4 km^2 a year. However, in recent decades the construction of levees, the damming of many of the delta's distributaries and the diversion of channels have reduced the supply of both sediment and fresh water to many wetland areas, accelerating erosion which is now proceeding at a rate of up to 100 km^2 a year (Day et al., 1993).

This highlights the need for a change of approach to the management of sensitive coastal areas. Bangladesh's recent Flood Action Plan has been criticised for its emphasis on 'hard' engineering solutions, attempting to stop floods rather than trying to control them. Instead of the traditional American emphasis on large-scale engineering works as a solution to flood hazard problems, an approach which is very expensive and which could not be afforded by most countries, there is a need to use the dynamic processes of environmental systems to man's advantage, to aid rather than obstruct the ability of natural systems to respond to sea level rise. Long-term solutions are required rather than short-term stopgaps, such as installing gravity drainage systems which will become inoperable if sea level rises.

IMPACTS IN TROPICAL AND SUBTROPICAL AREAS

For countries such as Australia there are fears that warmer sea conditions could bring tropical cyclones into areas where they have not previously occurred as well as increasing their intensity. One projection for Australia is that a 2°C rise in sea surface temperature could shift the margins of the cyclone belt from its current limits of 25 degrees south to as far as 31 degrees, making cities like Brisbane vulnerable to the associated storm surges and a much higher threat of flooding than formerly. Recent sea level rises and fears of further flooding have already begun to affect the inhabitants of some coastal areas. In a suburb of Redcliffe, near Brisbane, residents of a high-cost waterfront development are already concerned about the flooding of their gardens. An 80 cm rise in mean sea level would bring the average water level halfway up their gardens and storm surges would flood the houses. One or two residents have already moved out, citing as a reason the belief that either global warming would cause a rise in sea level, which would flood the properties, or that publicity about the greenhouse effect would make the houses unsaleable even if sea level remained unchanged. Estate agents in the area have been working hard to assure prospective buyers that global warming will not occur and that sea levels will not be affected. On what basis these claims are made is not clear!

More vulnerable still are island groups in the Indian and Pacific Oceans, though recent more modest estimates of sea level rise suggest that future scenarios for such areas, though involving considerable change, may not be quite as catastrophic as originally forecast. The Maldives, for instance, comprise over 1,200 islands with most of the land less than 3 m above present mean sea level. Some of the main islands are protected by coral reef barriers. Small rises in mean sea level could submerge these and expose the islands behind to erosion. The problems of the Maldives have been worsened by the quarrying of coral offshore for construction work linked to the developing tourist industry, while pollution has killed the coral in some areas. There is a need to restrict the areas of coral extraction to more remote uninhabited islands where it can do least damage, rather than allowing it on the larger islands close to densely populated areas.

Coral reefs can grow in response to sea level rises, though, like so many indices, the measurement of growth rates of coral reefs is far more complex and difficult than it first seems. A wide variety of growth rates have been calculated from different reefs ranging from 1 mm per year to as much as 13–14 mm per year. However, such figures are very site-specific and do not necessarily reflect wider trends, even within a single reef. The highest measured growth rates are associated with fringing and barrier reefs in areas such as north-east Australia and Hawaii (Fig. 5.10). Mid-ocean atolls show lower rates of growth, usually 1–3 mm per year, though sometimes as high as 5–8 mm. Such figures suggest that most reefs should be able to keep pace with projected rises of sea level. Reefs only appear to fail to keep up when sea level rise exceeds around 20 mm per year, rates which only occurred in the past during rapid deglaciation. However, at more modest rates of sea level rise there may be a time lag between a reef initially being drowned and the coral responding with renewed growth. During this time the reef may be vulnerable to damage or even destruction. When sea levels rise at 8–10 mm per year – a rate which is closer to some of the predictions for future sea level rise – it has been shown that reefs have trouble in responding.

The ability of reefs to keep pace with sea level rise will also be influenced by the frequency and intensity of tropical cyclones; increased storm activity may prevent coral growth, as may higher ocean temperatures. A severe hurricane can destroy the results of fifty years of coral growth. Any increase in storm intensity or the area affected by tropical cyclones might have significant impacts on coral growth.

Coral reefs flourish where sea surface temperatures do not fall below 18°C or above 28°C and within a certain range of salinity. High sea temperatures can cause the 'bleaching' of coral; thermal stress which can cause the coral to stop growing or kill it entirely if the temperature rises are severe and prolonged. A rise in sea temperatures of only 1–2°C for a period of weeks can cause temporary damage. Rises of 4°C above the norm, even for only short periods, can have more severe and long-lasting

Fig. 5.10. Fringing coral reefs off the Queensland coast. These reefs are relatively fast growing and should be able to keep pace with foreseeable rising sea levels. (Photo: Dr. A. Pringle)

effects. The 1982–3 ENSO event caused a rise in sea water temperatures of 2–4°C around Indonesia and the Galapagos Islands causing serious coral mortality. Other severe bleaching episodes have been reported from the Great Barrier Reef and the Caribbean. Some scientists consider these to be the first strong signal of the anthropogenic greenhouse effect in tropical waters. Coral reefs around Jamaica have already been devastated by a combination of overfishing, mining, excess inputs of sediment and nutrients due to pollution and, on top of this, extensive hurricane damage in the 1980s after nearly forty years without major storms (Hughes, 1994).

Another possible impact of global warming on sea level is the short-term effect of storm surges if there is a change in the area affected by tropical cyclones, the most serious of natural hazards which the inhabitants of coastal areas in the tropics have to face (Pittock and Flather, 1993 and Raper, 1993). The energy which drives tropical storms is derived from ocean heat so that there is a close relationship between sea surface temperatures and the occurrence of hurricanes. Sea surface temperatures of 26–27°C or higher are needed to generate tropical cyclones. Superficially, global warming might be expected to increase the areas of ocean with surface temperatures in excess of this threshold. An increase of frequency of cyclones of 37–63 per cent for a 1°C rise in sea surface temperature has been suggested on the basis of simple modelling exercises. However, warm sea surface temperatures are not the only prerequisite for hurricane generation; particular sets of atmospheric conditions are also needed and it is not clear how the occurrence of these might be affected by global warming. Empirical evidence indicates that the frequency of tropical storms was reduced during the cooler phases of the Little Ice Age, while the warm 1980s have seen an increase in the frequency of storms in the Caribbean and South Pacific, compared with the early twentieth century when sea surface temperatures were lower.

THE ANTARCTIC ICE SHEETS

The scenarios which we have considered, associated with moderate sea level rises, may sound bad enough but much more dire predictions have been put forward. These are based on the large-scale melting of the West and East Antarctic ice sheets (Mercer, 1978 and Thomas et al., 1979). The entire Antarctic ice cover would, if melted, raise sea levels by around 65 m. Antarctic ice comprises two main elements. The more stable East Antarctic ice sheet is a huge dome over 4 km thick. The West Antarctic ice sheet is made up of three smaller domes grounded on bedrock below sea level in many places. It is drained by ice streams, many of which flow into ice shelves, floating sheets of ice in huge bays, moving forward several hundred metres a year. It has been suggested that this situation is very unstable. A large area of the ice sheet is grounded on the ocean floor below sea level with an extensive floating ice shelf which is pinned in several places where it runs aground. The floating ice shelf buttresses the ice sheet behind. The fear is that warmer conditions would accelerate melting at the base of the ice shelf causing it to thaw and reducing the back stress on the ice sheet which might then surge forward and calve into the sea. There are precedents for this in the past including the final collapse of the Laurentide ice sheet over what is now Hudson Bay 7–8,000 years ago. The East Antarctic ice sheet is widely considered to be much more stable; it has generated a climate so cold that it is considered impervious to any foreseeable temperature rises. More recently this idea has been challenged and it has been suggested that the East Antarctic ice sheet has varied more in size in the past and may be more susceptible to climatic change.

The West Antarctic ice sheet would, if totally melted, boost sea levels by around 5 m. Early studies suggested that this could happen relatively quickly, producing a rapid, catastrophic sea level rise (Mercer, 1978 and Thomas et al., 1979) but more recent work suggests that, with the projected rates of global warming, a time-scale of centuries or longer would be needed for significant change to occur, though it is possible that thinning of the ice shelves on a time-scale of decades might set in motion a positive feedback mechanism which would lead to irreversible changes in the ice sheet by large-scale melting in the long term. The possibility of this ice sheet breaking up and melting on a time-scale of decades is no longer, however, taken seriously by most scientists.

One problem with assessing the vulnerability of these ice sheets is that our knowledge of the characteristics and behaviour of the Antarctic ice cap is still very limited. The rate at which the West Antarctic ice sheet might lose ice by calving is determined by the topography of the sea-bed as calving is increased by the depth of water. This is still not known about in any detail. The possibility of a modest rise in sea level triggering off massive calving and a surge of the ice sheet cannot be completely dismissed. The feedback mechanisms within both the Greenland and the Antarctic ice sheets are complex and far from well understood. For instance, warmer conditions in the Arctic are likely to increase the volume of the north Greenland ice cap due to additional snowfall while, at the same time, decreasing the volume of the southern part of the ice sheet due to greater melting. On a smaller scale, within southern Greenland, increased melting would be likely to be concentrated around the margins of the ice sheet while increased accumulation thickened the interior ice. This would then steepen the slope of the ice sheet and affect the speed of ice flow. Increasingly sophisticated models of ice sheet behaviour are aiding the understanding of the dynamics of the Greenland and Antarctic ice sheets but much remains uncertain (Budd et al., 1987 and Oerlemans, 1993b). A recent suggestion is that the distribution and character of the till beneath the ice sheet may have been responsible for periodic surges in the past. The condition of the till is not directly related to contemporary climate so that collapse might not necessarily occur under warmer conditions (MacAyeal, 1992).

RESPONDING TO SEA LEVEL RISE

We have already seen how complex are the processes operating in coastal environments and how profound is the scale of existing human impact in many areas. It is important to appreciate that many coastal areas are already under threat of flooding. Sea level rises will only exacerbate an existing problem. An effective response to the threats posed by sea level rise would be to evaluate existing coastal management strategies and alter them so as to make it easier for natural processes to cope with higher sea levels; in the Maldives, stopping coral extraction may be more effective protection than building sea walls.

Responses to sea level rise include the evacuation of threatened areas by moving human activities and developments further inland, trying to maintain the existing coastline by defensive works, and seizing the initiative by enclosing and reclaiming coastal areas. Such aggressive high-cost options requiring large-scale engineering works are likely to be favoured mainly in highly urbanised countries. The cost of such measures, though substantial, would only represent a tiny proportion of gross national product. However, for a poor country such as the Maldives the IPCC has estimated that the annual cost of preventing a 1 m rise in sea level could take up 34 per cent of GNP.

Substantial lengths of coastline in developed countries are already protected – 83 per cent of the coast of Belgium for instance has some kind of sea defence protection (Tooley and Jelgersma, 1992) – but in many other areas coastal protection is barely sufficient to meet current threats and would require expensive rebuilding to meet the challenge of rising sea levels. Abandonment of land may be the most economic strategy for most of the world's coasts outside the most developed areas. This includes not only the coastlines of underdeveloped countries but many of the less developed parts of coastlines such as those of the USA. This is already occurring in areas of the Nile delta which are being eroded and in eastern England where progressive cliff erosion is occurring.

There are problems for coastal wetland conservation sites, many of which are already threatened by changes in river hydrology. On the other hand, a retreat from the coast would create new wetland areas on former agricultural land. In areas like the Maldives and parts of south and south-east Asia and the Pacific islands the resettling of people displaced from coastal areas may become a major problem. On some Pacific islands the first refugees from rising sea levels have already been reported, a trickle which could, in a few decades, become a steady flow. But in practice it may be difficult to distinguish 'global warming' refugees from other displaced people as sea level is only one factor which threatens the existence of residents of coastal areas in the poorer countries of the world.

In Britain the latest flood control and coastal defence policy statement from the Ministry of Agriculture advocates an approach where natural coastal processes are adapted and supplemented using 'soft' rather than 'hard' engineering options. Moving defences landwards to higher ground, a policy of managed retreat, is far cheaper than building concrete sea walls, while abandoning existing defences would create additional wetland areas which would help protect the new defences. In sample tests, constructing new inner embankments has been shown to cost only a third of the price of strengthening existing defences (Arnell et al., 1994). However, if such a policy was implemented more widely provision would have to be made to compensate landowners for their losses and there is concern that such a policy might be supported by governments primarily as a cost-cutting exercise (Fig. 5.11).

In areas which have been developed, especially urban ones, it is more likely that efforts will be made to maintain the present coastline in the face of rising sea levels. In The Netherlands, where about half the coastline is already artificial, the plan is to hold the present coastline by height-

Fig. 5.11 Salt marshes like these at Silverdale, Lancashire, would be vulnerable to erosion with only a modest rise in mean sea level. (Photo: I. Whyte)

ening and improving existing defences (Bird, 1993). This is obviously a high-cost option, only feasible for the most densely populated coastal areas. For Hamburg, the expenditure on improved sea defences over the next twenty-five years has been estimated as DM3–4,000 million – for a single city. As well as urban centres, coastal nuclear power stations will also require protection (Tooley and Jelgersma, 1992). In tourist areas beach erosion can be substantially reduced by constructing breakwaters offshore, something which is already taking place in some resorts. But, where hotels are built close behind the beach, more substantial protection may be needed, and artificial replenishment of beaches with sand is likely to become more common. Several hundred beaches in the USA are already treated in this way. As the costs of such actions increase it may become necessary to concentrate on protecting particular beaches in resort areas. Such high-cost options are only feasible to a limited extent in key areas of developing countries. Flood protection barriers near major cities, such as London's Thames Barrier, may become more common. One is already under construction to protect Venice. Sea defences will be crucial in protecting cities like Hong Kong and Bangkok where subsidence is adding to the problem of eustatic sea level rises.

If new coastal defences are to be built, a more aggressive strategy is to construct them offshore and reclaim land to help offset construction costs. In some areas this could be done by means of tidal barrages generating renewable energy. In the case of bays such as Port Phillip Bay in Australia, which has a 260 km coastline but only a 3 km-wide entrance, such an approach would be cheaper.

More careful environmental and land use planning in coastal areas is needed. This involves careful assessment of which areas are most at risk, the production of plans to reduce their vulnerability, limiting developments in coastal areas and preparing for emergencies caused by extreme weather events, such as storm surges, as well as the effects of gradual rises in mean sea level. Data on related processes such as river discharges, sediment loads and groundwater reserves would also be essential. Coastal lowlands can then be zoned on the basis of levels of risk. Areas of high risk are linked with high tidal ranges, frequent onshore winds and the occurrence of storm surges. The risk is higher where existing natural and man-made sea defences are inadequate. Stricter controls over future developments in especially vulnerable areas could reduce the problems. Many areas in immediate proximity to the coast have already been developed for residential, industrial and recreational purposes. There is an urgent need to control and, if necessary, restrict further activities of this kind in areas of highest risk.

A final response is to attempt to reduce the rate of sea level rise. This can be done by cutting emissions (Chapters 4 and 7) but also through large-scale interference with the hydrological cycle (Newton and Fairbridge, 1986). Pumping of seawater into underground aquifers or low-lying basins such as the Qattara Depression in Egypt would abstract water from the oceans, offsetting some of the effects of thermal expansion and the melting of land ice. Such ideas may seem to be in the realms of science fiction but they are technically (though not necessarily economically) feasible. Their desirability from the point of view of interfering with regional ecosystems is another matter and leads us to the consideration of another series of impacts of climatic change.

6

The Impact of Global Warming on Natural Ecosystems and Agriculture

Climatic changes on the scale and speed considered in Chapter 4 would have major influences on natural ecosystems and agriculture. The impacts of climatic change on natural ecosystems are, in some ways, potentially even more serious because agriculture has the capacity to adapt faster to climatic change. In the past, with natural average temperature changes rarely occurring at a rate of more than 1°C per 1,000 years, ecosystems could keep up by altering their composition and boundaries. Studies of the response of vegetation to the temperature changes which occurred during the most rapid phases of postglacial warming have demonstrated that there were major time lags in the responses of ecosystems to climatic change, resulting in considerable ecosystem instability. The responses of natural ecosystems in the future would be further hindered as so much of the land surface has been altered by man. The speed of predicted climatic change could be too great for many species of plants and animals to cope with by migrating to new areas with suitable conditions. This could result in the extinction of many species and a decrease in biodiversity. On a wider scale there are fears that entire major ecosystems, such as the northern coniferous forests, might be virtually squeezed out by warmer conditions.

Studies of the impact of future climatic changes at a regional scale have mostly adopted a fairly simple approach. First they have compared climatic scenarios resulting from an instantaneous doubling of carbon dioxide levels with current conditions. However, as temperatures rise and precipitation patterns will occur progressively, regional economies are likely to change significantly in the period required for climatic changes to develop, altering their capability to adapt to new conditions. Second, the highly generalised output from GCMs leads to temperature and precipitation changes being applied uniformly across the large areas of GCM grid boxes. Third, studies of impacts usually focus on specific aspects of a region such as forests, hydrology or agriculture and, within agriculture, on particular crops. Finally, impact models have rarely tried to build in management responses to changing environmental conditions. A new generation of more detailed integrated models is being developed which looks at a whole range of interrelated aspects of particular regional economies. (Rosenberg, 1993). However, in order to survey a wide range of problems this chapter adopts a thematic approach using regional examples where appropriate.

CARBON DIOXIDE AND PLANT GROWTH

Rising carbon dioxide levels will not only affect climate, they will also have a direct, beneficial influence on plant growth. Increased carbon dioxide acts as a fertilizer for many types of plant (Parry, 1990). Plants use chlorophyll to capture carbon dioxide from the atmosphere, converting it into plant tissue and energy-rich compounds such as starch and sugar. More carbon dioxide in the atmosphere should increase the rate of photosynthesis at levels of carbon dioxide up to four times the pre-industrial concentration, or around 1,000 ppmv. As more carbon dioxide is taken up by plants, and more plant growth occurs, negative feedback is involved which will tend to slow the rate of increase of carbon dioxide in the atmosphere.

However, not all plants respond markedly to carbon dioxide rises. There are two main groups of plants with different systems for capturing carbon dioxide through photosynthesis. The first, C3 plants, are so designated because the first product in the sequence of biochemical reactions involved in photosynthesis has three carbon atoms. These plants include wheat, rice, barley, oats, potatoes, sugar beet, soya beans, sweet potatoes, cassava, bananas, coconuts, chickpeas and most fruits and vegetables. They react positively to increased carbon dioxide levels leading to greater growth and increased yields, possibly between 10 per cent and 50 per cent as long as sufficient water and nutrients are available.

The second group, C4 plants, in which the first product has four carbon atoms, is less responsive to increased carbon dioxide levels. Improvements in yields range from zero to around 10 per cent. C4 crops include maize, cane sugar, millet and sorghum. C4 crops are mainly tropical and are grown most widely in Africa. Many of the world's major pasture grasses are also C4 plants, so the carrying capacity of the world's major grasslands is not likely to benefit significantly from carbon dioxide fertilization.

C4 plants account in total for only about 20 per cent of world food production, 14 per cent being maize. Maize is nevertheless important as it makes up around 75 per cent of all traded grain and is the major grain used in food aid for famine-hit areas. C3 plants, as well as benefiting from carbon dioxide fertilization, may suffer less competition from weeds, many of which are C4 plants. On a global scale, food supplies depend on around twenty crops, sixteen of which are C3. Fourteen of the world's seventeen most troublesome weeds are C4 plants growing in C3 crops. C4 weeds may become less competitive, while C3 weeds in C4 crops could become a greater problem.

All this sounds as if the carbon dioxide fertilization effect will offset many of the problems which agriculture may face as a result of rising temperatures and potential moisture deficits; but it may be too good to be true. Experiments on the effects of atmospheric carbon dioxide increases on improving plant growth have generally been carried out under carefully controlled laboratory conditions and may not apply on a larger scale in the real world. Increases in crop yields may, for example, be temporary if plants adapt to higher levels of carbon dioxide. C3 plants may yield more with carbon dioxide fertilization but what about the quality of the harvest? Rises in carbon dioxide, while increasing the carbon content of plants, may reduce their nitrogen content, cutting protein levels and nutritional value. There are also worries that the potential of plants to respond to carbon dioxide fertilization may be restricted by insufficient moisture and lack of adequate nutrients in many soils and by competition for available nutrients from soil micro-organisms.

The gap between laboratory experiments and conditions in the field has been emphasised by recent work which enclosed an area of Alaskan tundra with a greenhouse to measure its response to changing carbon dioxide and temperature conditions (Oechel et al., 1994). It was found that the doubling of carbon dioxide on its own increased the rate of carbon dioxide fixation for only a few weeks. The likelihood is that, in an ecosystem like this, increased carbon storage could rapidly deplete nutrients and reduce the scope for further growth. The study found that, when a doubling of carbon dioxide was combined with a temperature rise of 4°C, carbon dioxide absorption continued for longer. However, the research was only carried out for a three-year period. It is possible that warmer conditions increased the rate of decomposition of soil organic matter, releasing more nutrients for plant growth, a process which again would eventually deplete the soil and cause further plant growth to be reduced.

Higher carbon dioxide levels may also affect the water balance of plants. Plants exchange gases with the atmosphere through openings, or stomata, in their leaves through which carbon dioxide is taken in and water vapour released. With higher carbon dioxide levels, both C3 and C4 plants should be able to capture the carbon dioxide they need more quickly so that the stomata stay open for shorter periods and less water is lost. With a doubling of carbon dioxide it has been suggested that transpiration and loss of water vapour by plants may be reduced by 23–46 per cent, but the effect of this improved water retention under real, as opposed to laboratory, conditions is not clear. Greater water use efficiency may help plants in areas where lack of moisture currently limits growth. That this mechanism is already starting to work is suggested by comparisons of the number of stomata of modern plant leaves with examples of the same species from 200-year-old botanical collections. The average numbers of stomata in temperate tree leaves have already fallen indicating that plants are responding to higher carbon dioxide levels.

There are indications that the balance between C3 and C4 plants may have varied in the past in line with changes in carbon dioxide levels. In New Mexico for instance a switch from C4 dominated grasslands to C3 dominated scrub at the end of the last glaciation coincided with a rapid rise in atmospheric carbon dioxide. Local temperature conditions appear to have remained relatively constant suggesting that the change in carbon dioxide concentrations rather than climate affected the vegetation (Cole and Curtis-Monger, 1994) A similar influence may lie behind the rapid expansion of scrublands over large areas of the American south-west within the last two centuries (Idso, 1992).

IMPACT ON ECOSYSTEMS

The study of possible future changes in ecosystems resulting from global warming can be carried out via computer modelling exercises, the study of analogues from the past, and the examination of changes occurring at the present. Changes in ecosystems can be modelled on a large scale by mapping the climatic variables which define the limits of the major vegetation zones and using GCMs to try and show how these limits might change under different climatic scenarios. Most GCMs agree in forecasting a marked drying in mid-latitude continental interiors in the northern hemisphere with a probable shift from forest to scrubland and grass. They also suggest that there would be a decrease in the area of tundra of over 30 per cent with an equivalent doubling of carbon dioxide as the limits of boreal forests advance northwards. A decrease in desert areas due to a more vigorous monsoon circulation is also a possibility. In modelling such changes, however, the scale of the time lag between climatic change and vegetation response must be remembered. The spread of vegetation would be delayed by the time required for plants to migrate and, in some cases, for suitable soils to develop.

Recent warming has already begun to produce changes. The spread of vegetation in the Antarctic Peninsula has already been mentioned. In the Austrian Alps scientists have discovered that, on summits over 3,000 m, there has been an increase in the species richness of plants in the last seventy to ninety years due to upward migration (Fig. 6.1). Some species have been increasing their upper altitudinal limit by up to 4 m per decade (Grabherr et al., 1994).

Fig. 6.1 Vegetation colonising bare ground in the Austrian Alps. Recent warming in this area has caused plants to advance to higher altitudes. (Photo: I. Whyte)

Models suggest that all the world's vegetation zones would be affected by the level of global warming thought likely to accompany the equivalent doubling of carbon dioxide. Desert and polar areas would probably be the most stable regions (Fig. 6.2) but because of the greater predicted warming in higher northern latitudes the boreal forests would be especially affected (Monserud et al., 1993). Because of this it is useful to focus on ecosystems in northern areas.

Recent changes in the Arctic highlight the potential for more drastic developments in future. In the basin of the Mackenzie River between the early nineteenth century and the 1940s, a 3°C rise in average temperatures caused permafrost to retreat by several hundred kilometres and the coniferous forest boundary to move north (Woodward, 1992). Further warming of 1°C in Canada or Russia could cause a shift in the permafrost boundary of 200–300 km and a 2°C rise 500–700 km. As noted in Chapter 2, the thawing of permafrost might release large quantities of methane from northern peat bogs, reinforcing global warming.

Arctic environments are especially sensitive to temperature change, particularly in view of the possibility that future warming may be greatest in high latitudes. One impact of warmer conditions would be the thickening of the active layer above permafrost and a thinning of the permafrost below, producing a marked retreat in the permafrost boundary. In the past, permafrost limits have shifted over considerable distances

Fig. 6.2 The edge of the Sahara desert. Desert margin ecosystems would be little affected by temperature rises likely to result from the man-enhanced effect but changes in precipitation could be much more significant. (Photo: Dr. P. Barker)

with comparatively slight changes in global mean temperatures. During the mid-Holocene optimum the southern limit of discontinuous permafrost in Russia was as much as 1,000 km further north than it is today. Warming from the late nineteenth century onwards in the Mackenzie River area of the Canadian arctic has pushed the permafrost boundary north by about 350 km. A 2°C increase in global mean temperatures, producing a 7–9°C rise in winter and 4–6°C in summer temperatures in the Arctic, could dramatically reduce the extent of permafrost within a few decades. Whole landscapes would be affected as permafrost melting occurred with the formation of thermokarst features as the surface of previously ice-laden ground collapsed. There would also be an increase in mass movement and slope instability which would have an impact on surface form, drainage patterns, and river sedimentation as well as creating problems for man-made structures and engineering works.

A much more finely tuned approach to the study of changes in ecosystems resulting from climatic shifts uses large-scale models of particular ecosystems which plot the growth of individual trees within small areas and monitor how different species react to changing climatic parameters (Woodward, 1992). Such models can show how the competitive advantage of different species might change under a new climatic regime but such high resolution modelling can only be applied to small areas because of data and computing limitations. It is difficult to generalise from them and they may not provide answers which are appropriate at larger scales due to complex variations in soil, drainage conditions, and micro-climate.

Effects on vegetation

How quickly would vegetation respond to climatic change? In North America, a 1°C rise in average annual temperatures would shift the theoretical forest boundary north by 100–125 km. For an equivalent doubling of carbon dioxide, predictions indicate that a shift of boreal forest limits 500–1,000 km northwards. Would the time-scale for real changes be one of decades, centuries or millennia though? The study of vegetation changes in the past may help us by providing concrete evidence, though under conditions which are rather different from those likely to occur in the future. The spread of vegetation may be delayed in some areas by lack of suitable soils. There are indications that, at the end of the last glaciation, the lack of developed soils may have slowed the spread of forests in Britain for 1,000 years after temperatures had risen to suitable levels. In early postglacial times in Europe, trees seem to have migrated at maximum rates of 0.02–2 km a year with a mean of around 0.4 km. In North America the process was slower.

Most trees would lag behind global warming by centuries or millennia on this basis. But these rates are derived from pollen evidence which may be misleading. Pollen diagrams may not pick up the original wave front of tree colonisation because the amount of pollen they produced was so small, presenting a misleading picture. Some studies of recent plant colonisation suggest that the response of vegetation may be much faster under certain circumstances (Pastor, 1993). Nevertheless, the rates at which carbon accumulates in soils – as low as 0.1–0.01 gm per m^2 per year – suggest that many existing soils must have taken thousands of years to reach full development (Woodward, 1992). Studies of shifts of vegetation during the Holocene have shown that the migration of trees during periods of climatic amelioration is faster than the recession during phases of climatic deterioration, because the retreat of the treeline is delayed until existing trees die off and are not replaced (Grove, 1988).

Forests are especially vulnerable to the effects of climatic change through temperature stress, changes in precipitation, increases in pests and diseases and competition from other ecosystems. It is easy to shift forest boundaries around on a map in response to higher temperatures but it is far harder in practice for trees to adapt to changing climatic conditions. After the last ice age, forest boundaries in the USA moved northwards in response to warmer conditions at between 20 and 40 km a century. In central Canada though there is evidence that the boundary between tundra and boreal forests has shifted quite rapidly in the past; black spruce migrated at rates of around 200 km a century following deglaciation (Pastor, 1993).

Concern regarding the impact of climatic change has centred on forests in temperate areas. The temperature rises that have been predicted would require forests to adapt at a much faster rate than at the end of the last ice age. Under such conditions it would be very difficult for the forests to keep up with changing conditions. At the southern boundaries of the forest zones, water stress would cause dieback. Species would tend to die out faster at the southern boundary of the forest belt than they would be able to spread at their northern limits. A 1°C rise in temperature might push the forest/grassland boundary northwards by 60–100 miles. Under warmer conditions air pollution and insect pests could slow the ability of trees to colonise new areas and speed their dieback in sensitive areas.

Recent years have seen some devastating forest fires in areas including eastern Australia and the western USA. Warmer conditions would increase the fire risk still further. In many parts of the central USA there is evidence of an increased fire frequency in the warmer, drier fifteenth and sixteenth centuries compared with the cooler conditions of the seventeenth to nineteenth centuries. The susceptibility of forests to fire damage under given sets of moisture conditions is likely to be greater in the future due to human activity (Wyman, 1991). Increased frequency of forest fires has been widely predicted as an accompaniment to hotter, drier climatic regimes in areas such as the western USA. The incidence of fires in Yellowstone National Park has increased since the end of the 1970s after a period from around 1940 with few fires. In part this has been due to efforts at fire suppression but the fire of 1988 destroyed nearly 400,000 ha; no other fire in the present century has destroyed more than 9,000 ha (Balling et al., 1992), suggesting that the fire risk is increasing. The poor quality of information on changes in wind speeds in current GCMs is one of the problems which makes an assessment of fire risk under possible future climatic conditions difficult (Torn and Fried, 1992).

There is growing evidence that, under present conditions, northern coniferous forests are an important sink for carbon dioxide (Fig. 6.3). The ability of these forests to continue absorbing carbon dioxide is already threatened by timber extraction and acid rain and may be further hit by temperature rises. In Siberia, which contains 57 per cent of the world's coniferous forests, large areas of timber are currently being destroyed in order to provide hard labour for convicts. Much of it is simply left to rot. In upland Britain, however, where soil moisture is not a limiting factor on tree growth, a temperature increase in summer from 14°C to 17°C is likely to increase the rate of breakdown of litter and nutrient release by 10–30 per cent and benefit tree growth (Cannell and Hooper, 1991).

There are implications for commercial forestry too. In Britain, planted conifers such as Sitka spruce require certain minimum temperatures to thrive and might suffer severely with warmer climates. In addition conifers in many upland plantations in Britain are growing near the limits of their wind tolerance. The gales of 1987 and 1990 showed the vast amount of damage which could be done in future if conditions became more stormy.

Changes in forest ecocystems due to global warming will not just affect

Fig. 6.3 The northern margins of the coniferous forest zone in Sweden. The importance of these forests as a sink for carbon dioxide may have been underestimated in the past. (Photo: Dr. P. Vincent)

their boundaries but will also involve alterations in species composition which may take centuries to settle down to a new equilibrium state. Recent studies of pollen from the now largely cleared mixed forests of South Ontario show that from around AD1400 the dominant warmth-loving beech trees were replaced by oak and then pine (Campbell and McAndrews, 1993). It is thought that this was due to the onset of cooler conditions with the start of the Little Ice Age. The changes have been simulated in computer models which suggest that, had the forests not been cleared, the ecological changes brought about by Little Ice Age cooling would still not have reached an equilibrium state 600 years later. In the model, simulated cooling similar in scale to that of the Little Ice Age caused a decline in beech which opened out the canopy, allowing other species such as oak and pine to expand, but reducing the biomass of the forest. The implications of this are that rapid climatic change could affect forest biomass by causing large-scale tree death which would add to atmospheric carbon dioxide.

It is also worth remembering that the changing forest boundaries would themselves have an effect on climate through alteration of the surface albedo. In the past it was thought that the extent and character of coniferous forests was dependent on climate. Now it is realised that there is a dynamic equilibrium between the two (Bonan et al., 1992). Boreal forests increase winter and summer temperatures due to their absorbtion of radiation. Modelling of the impact of the northward expansion of boreal forests during the mid-Holocene optimum has suggested that the resulting reduction of albedo could have been sufficient to warm high latitudes by up to 4°C in spring and 1°C at other times of the year. Thus the northward expansion of the coniferous forest belt could further enhance warming in the Arctic (Foley et al., 1994).

Predictions of temperature and especially precipitation changes in tropical areas are not certain enough to provide an indication of how rain forests might be affected. The main worries about forests in tropical areas concern the effects of deforestation due to human action and the possible effects that this might have in reinforcing global warming rather than the impact of global warming on the forests. It should be noted, however, that modelling studies have suggested that, for the Amazon rain forest, large-scale deforestation could lead to a regional-scale drying of the climate which could imperil surviving, protected rain forest areas.

Effects on animals

So far the focus has been on plants but animal species may also be severely affected if global warming occurs as predicted. Polar bears, for instance, would be vulnerable if there was a widespread reduction in the area of arctic sea ice, for the bears depend on hunting seals from the ice in winter. Many species of birds whose distribution is tied to specific ecosystems could also be threatened. In western Europe, for instance, even a modest rise in sea level could inundate the salt marshes which form the wintering grounds for large numbers of migrant waders and the nesting areas of many resident species. A report by the Institute of Terrestrial Ecology in 1989 suggested that, under such conditions, common British species such as the redshank could become rare (Cannell and Hooper, 1991).

In contrast, warmer conditions would encourage some birds to extend their present range. A number of species spread northwards during the first half of the twentieth century as a response to warmer conditions. Black-headed gulls, swallows, starlings and fieldfares have moved into areas like the Faeroes, Iceland and Greenland, while Mediterranean species like the serin and Cetti's warbler have pushed into western Europe (Grove, 1988). Some species of butterfly such as the white admiral and comma have also extended their ranges considerably within Britain during the past century as a result of warmer conditions (Wyman, 1991). Individual species are likely to respond to changing climatic conditions at different rates and in different ways, breaking up the structure of existing ecosystems and creating new assemblages of plants and animals which may have no present-day analogues. This will expose many species to new, exotic competitors.

The incidence of short-term extreme conditions may have more influence on the survival of many species than gradual shifts in means. Many species with limited distributions which are already under threat could be wiped out by a year or two of anomalous conditions such as droughts or floods. One of the world's rarest wildfowl, the Laysan teal, which inhabits a small, remote island in the Hawaiian group has recently had its total population cut from around 500 to under forty due to an unusually severe drought. The efficiency with which species can reproduce and migrate will obviously influence their success under changing climatic conditions. Species are more likely to become extinct if they inhabit limited geographical ranges. The ability of many species to shift in response to changing climatic conditions is likely to be hampered by the fragmentation of their present habitats and because of stresses on their reproductive and dispersal abilities caused by influences such as acid rain and introduced insect and fungal pests. In the case of plants, migration may be retarded by the reduction or even elimination of species of animals and birds which formerly aided the dispersal of their seeds (Wyman, 1991).

Animals may be capable of great mobility but this may be of little benefit if their distribution is limited by the range of particular food supplies. Loss of habitat due to growing population and increased pressure on the environment makes species even more vulnerable if remnant populations become confined to small islands or refuges in the middle of agricultural or urbanised areas with only narrow corridors along which migration to suitable new habitats could occur, unless man can assist in the process of re-colonisation. Animals and birds which migrate over long distances may be particularly affected by changes in the habitats of stop-over zones used during migration as well as their breeding and wintering areas. Indeed, a combination of climatic change, habitat loss and pollution may underlie the substantial decline in numbers of some species of American shorebirds in recent years (Wyman, 1991). Changes in temperature may alter peaks of food abundance in habitats so that they are out of phase with migration times. However, man rather than climate has been the most important influence in changing bird populations and their distribution for many centuries. It is likely that changing climatic conditions in the future would affect birds primarily through their impact on agriculture and forestry along with habitat changes caused by rising sea levels and the coastal management responses to such rises.

TROPICAL RAIN FORESTS AND CLIMATIC CHANGE

We have seen that climatic change is linked with a range of other environmental problems. One which is frequently cited is the destruction of tropical rain forests. This is the largest change in land cover currently occurring on Earth. Northern coniferous forests are not being destroyed on the same scale, though the amount of large-scale commercial timber extraction in these forests looks set to grow substantially in the future. In temperate areas, deforestation has occurred throughout historic and prehistoric times. The average rate at which forests are being cleared has fallen in the last century and regrowth has been occurring in many areas, acting as a net carbon sink. Afforestation has greatly increased the area of woodland in Britain during the last seventy-five years, for example. In the Mediterranean the abandonment of much marginal agricultural land has led to substantial expansion of woodland and scrub (Jeftic et al., 1992).

In recent years the destruction of tropical rain forests has attracted increasing attention not merely because of the direct damage to the world's most complex, richest and least-known ecosystem, but also due to its potential impact on global climate. The vegetation and soils of natural or semi-natural forests can hold twenty to one hundred times more carbon per unit area than agricultural land. Carbon dioxide is released from burning vegetation, from the decay of felled trees and cleared scrub and from loss of carbon in former forest soils. Between 1860 and 1980 deforestation and other land use changes added between 80 and 150 Gt of carbon to the atmosphere, about a third of the total carbon dioxide released by human activities. There is a wide margin of error in these figures due to a lack of knowledge of the extent of forests in the past and the rate at which they have been converted into agricultural land. Given the bad press which countries like Brazil have received in recent years it is worth remembering that much of the carbon dioxide produced by deforestation since the mid-nineteenth century has come from developed countries such as the USA. Temperate areas provided the main source of carbon dioxide from vegetation changes in the late nineteenth and early twentieth centuries but, since the 1950s, tropical forests have taken over, releasing two or three times as much carbon as forests in mid- and high latitudes in the 1980s.

The background of rain forest destruction is well known; until the 1960s farming in rain forest areas was mainly small-scale slash and burn shifting cultivation which was capable of making sustainable use of the forests provided that numbers of cultivators were not too great. Since then rain forests have been increasingly opened up to larger-scale commercial exploitation, often by farmers driven out of other areas as a result of environmental degradation caused by poor land management; shifted rather than shifting cultivators. Large-scale timber extraction and mineral exploitation have also affected the area of rain forests.

The total area of tropical rain forest is hard to calculate accurately but may be around 7.8 million km^2. There are considerable problems in estimating the area of deforestation though; most figures are based on satellite imagery and it is hard to be certain about how much forest has been degraded rather than totally destroyed. Early work based on remote sensing estimated a loss of 8 million ha a year from Brazil but this has now been considered a substantial overestimate. In the late 1970s it is thought that about 75,000 km^2 of rain forest were being cleared each year worldwide. By the late 1980s this had risen to around 142,000. Brazil had in the region of 2.9 million km^2 of rain forest in the 1960s, or 3.6 million if areas of transitional forest are included. Today this has been reduced to about 2.2 million km^2, accounting for around 28 per cent of the world's rain forests. For the whole of the Brazilian rain forest, total losses by 1975 were only estimated at around 29,000 km^2 but by 1988 400,000, or 11 per cent of the original area, had been destroyed, nearly 70 per cent of the losses occurring since 1980.

Brazil is currently estimated to be losing 50,000 km^2 each year due to clearance for agriculture, cattle ranching, timber extraction, energy production and mineral exploitation. The rate of deforestation in Brazil averaged 2.1 million ha 1978–89, falling to 1.4 million ha in 1990. Deforestation may have peaked in the late 1980s but it is still substantial. Deforestation in Brazil seems likely to continue and even increase due to government plans for new settlement in many rain forest areas and the continuing exploitation of minerals, oil and natural gas, with plans to flood some forest areas to generate hydroelectric power.

While Brazil has attracted most attention, the rate of rain forest destruction in some other countries has been higher. Nigeria is losing its rain forest at an estimated rate of 14 per cent a year against 2.3 per cent for Brazil. Indonesia and Zaire, which both have around 1 million km^2 of rain forest, a significant amount of the world total, are both expanding their logging programmes, not to mention inroads due to agriculture. At current rates of destruction, by 2000 many south-east Asian countries like Thailand, Vietnam and Burma are likely to have lost most, if not all, of their primary rain forest. A similar position is likely in many parts of Central and South America by this date with little rain forest left in Mexico, and that of Equador, Peru and Bolivia being heavily depleted.

In recent years the burning of rain forests is estimated to have contributed between 15 and 30 per cent of the carbon dioxide added to the atmosphere by man. The estimated carbon release from deforestation, mainly in the tropics, ranges from 0.3–1.7 Gt per year to 1–2.6 Gt, including 0.2–0.9 released from soils. Deforestation also releases significant amounts of methane and nitrous oxide so that its total contribution to current greenhouse forcing may be in the order of 20 per cent. However, given the rate at which rain forests are likely to be exhausted at current rates of removal, the total contribution of tropical deforestation to the man-enhanced greenhouse effect by 2100 will be much smaller. Total removal of all rain forests would add only another 35–60 ppmv of carbon dioxide to the atmosphere. Compared with releases from the combustion

(a)
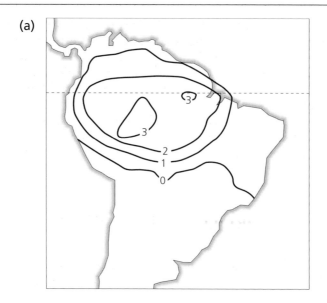

(b)
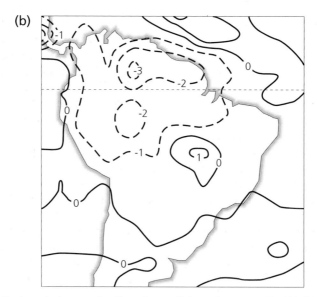

Fig. 6.4 Projected changes in climatic conditions due to removal of tropical rain forests from South America. A: annual mean surface temperature (°C). B: precipitation (mm per day – negative contours shown by dashes). (After Houghton et al., 1990)

of fossil fuels this is relatively limited. Calling a halt to rain forest destruction is highly desirable on ecological grounds to preserve as much species diversity as possible but, even if the rate of deforestation was to be drastically reduced within the next few years, the contribution towards reducing greenhouse forcing and future temperature rises would be limited. At a purely regional level, however, continuing deforestation is likely to have a serious impact on climate. Deforestation increases surface albedo leading to greater aridity. A 10 per cent increase in albedo resulting from large-scale removal of the Amazon rain forest might reduce precipitation in the region by 20–40 per cent (Fig. 6.4).

HYDROLOGICAL IMPACTS

Apart from the saline intrusion into estuaries and aquifers, considered in Chapter 5, climatic change is likely to have major impacts on water resources in terms of potential changes in both supply and demand (Arnell et al., 1994). A warmer world might, overall, be a wetter one but the distribution of precipitation is likely to change quite markedly in some areas, producing problems for water supply for agriculture, industry, energy generation, domestic consumption and navigation. Global warming is likely to influence a range of hydrological variables including atmospheric humidity, evapotranspiration, soil moisture, surface run-off, groundwater recharge, snowfall and snowmelt, river flows and lake levels, and the occurrence of droughts and floods. Changes in these will have impacts on water supply, distribution and use (Wyman, 1991). Changing hydrological conditions would also affect many aspects of soils such as rates of leaching and erosion.

In a warmer world increased evapotranspiration could lead to diminished soil moisture and stream run-off even if precipitation increases. Droughts in sub-Saharan Africa have become familiar since the 1960s. An increase in the frequency of ENSO events since the early 1980s has brought more frequent drought to Australia and parts of Indonesia. In late 1994, Japan was suffering from the worst drought in living memory with some factories being forced to close for lack of water and some reservoirs down to 10 per cent of normal capacity. The central USA and the Mediterranean are among other areas which have experienced recent severe droughts (see below).

In the past the problem of water supply planners has often been to forecast future demand accurately. In future the problem may be that of how to predict levels of supply. Greater variability of climate could increase the incidence of major floods as well as serious droughts. A major potential problem is the increased rates of evaporation that are likely to occur with even slightly higher temperatures, and the disproportionately great effects they might have on stream discharges. In the western USA, for example, an increase of temperature of 2°C with no change in precipitation might cause a 20 per cent drop in run-off.

Better perspectives on possible changes in river regimes can be obtained from long-term chronologies of the occurrence of major floods derived from geological evidence. A recent one, extending back some 7,000 years, has been constructed for some of the tributaries of the Mississippi (Knox, 1993). During a relatively warm, dry period 5,000–3,300 years ago major floods were rare. After that cooler, wetter conditions set in with an abrupt increase in flooding. Even larger floods occurred between AD1250 and 1450 in the transition from the medieval optimum to the Little Ice Age. The significant point about this chronology is that the marked changes in flood incidence occurred with changes in mean annual temperature of only 1–2°C and variations in precipitation of around 10–20 per cent. Human regulation of rivers and their basins tends to increase the severity of both flood peaks and periods of low water, enhancing any natural changes in the hydrological regime. In 1988 the Mississippi was dangerously low but, in 1993, floods covered an area the size of England spread through eight states, causing US$10 billion-worth of property damage and nearly overtopping flood defences around St Louis which were designed to stop all but a once-in-500-years flood.

In those areas where much of the available water is released from melting snow temperature rises, even with no change in precipitation, may increase the rate of spring snowmelt and alter the seasonal pattern of streamflow, producing higher spring peaks but reduced supplies later in summer. In many areas existing hydrological resources are already under severe pressure from increasing, competing demands and diminishing supplies. This aspect of global warming has attracted considerable attention in the USA where supplies of water are often barely adequate to meet current needs. About 40 per cent of streamflow is withdrawn for irrigation, supplying only 10 per cent of the arable area but accounting for nearly a third of the total value of crops. Power stations in the USA withdraw almost as much water as irrigation but the bulk of this is returned to river systems fairly quickly. Domestic use accounts for only about 10 per cent of water withdrawn but demands have increased rapidly in the last two decades.

On the basis of various hydrological features, it is possible to identify catchments which are most at risk from climatic change (Fig. 6.5). These include basins with low water storage capacity and high water consumption in relation to their average discharge, ones which generate substantial amounts of hydroelectric power and areas where there is a high dependency on groundwater supplies. This, and the variability of streamflow under present conditions, highlights many catchments in the west and south-west USA as being especially vulnerable, basins in the east and north-east less so. California, where a high proportion of agricultural

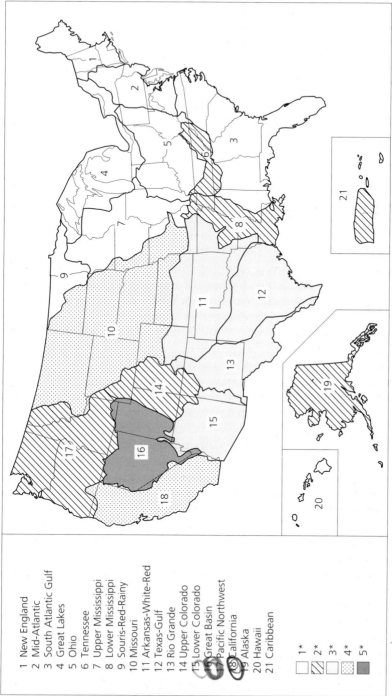

1 New England
2 Mid-Atlantic
3 South Atlantic Gulf
4 Great Lakes
5 Ohio
6 Tennessee
7 Upper Mississippi
8 Lower Mississippi
9 Souris-Red-Rainy
10 Missouri
11 Arkansas-White-Red
12 Texas-Gulf
13 Rio Grande
14 Upper Colorado
15 Lower Colorado
16 Great Basin
17 Pacific Northwest
18 California
19 Alaska
20 Hawaii
21 Caribbean

1*
2*
3*
4*
5*

Fig. 6.5 The hydrological regions of the USA, rated according to five important indices of water resource vulnerability; supply, demand, dependence on hydroelectricity, overpumping of groundwater and hydrological variability. White regions are vulnerable to one of these measures, shaded ones to all five. (After Mintzer, 1992)

output depends on irrigation and there is a large urban population, has potentially severe problems and major efforts are already under way to conserve water (Berk et al., 1993). Ambitious plans have been put forward to divert large volumes of water from rivers in western Canada including the Fraser, Liard, Skeena and Yukon, to supply southern California. The combined discharge of these rivers, when snow melts in spring, is equal to about 10 per cent of all the world's river water. The plan for a 1920 km channel and over 200 large dams would take at least thirty years to complete and has produced horrified reactions from environmental groups on account of its impact on one of the least spoiled areas in the world. There is also a worry that warmer conditions and higher evaporation would lower the level of the Great Lakes, perhaps by between 0.5 and 2.5 m, causing problems for navigation.

The Mediterranean is another area where existing problems of water supply may worsen with warmer conditions. In the late 1980s and early 1990s there has been a major drought which could be a foretaste of the future as GCMs suggest that, with higher temperatures, the northern limits of the Sahara may effectively jump the Mediterranean to reach areas like south-east Spain and Sicily. Growing tourist numbers and increasing demands for irrigation water are already pushing available hydrological resources to their limits. If drought in the Mediterranean intensifies, pressure to provide more water for irrigation is likely to increase. One answer is to improve water management by constructing dams but this has impacts on other aspects of the environment. A series of dams on the River Guadalquivir has lowered water tables in the Coto Doñana National Park, a wetland area of international importance at the river's delta, threatening wildlife already under pressure from local agriculture and tourist developments. In Greece, where water is scarcer still, even more ambitious schemes for diverting streams are under way, also posing threats to coastal wetlands (Jeftic et al., 1992). In Libya, where coastal aquifers have been exhausted, grandiose plans exist for pumping fossil ground-water from beneath the Sahara for hundreds of miles to the coast.

In Britain the droughts of 1975–6 and 1988–92 have shown that we can no longer take the availability of adequate water supplies for granted. Although model scenarios suggest that precipitation may increase during the winter months the trend for the summer is less clear (Chapter 4) and higher temperatures are likely to increase evaporation and loss of soil moisture. Increased winter precipitation may lead to more frequent and severe flooding (Arnell et al., 1994). It is not clear whether a shortening of the recharge season for groundwater supplies with drier summers would be compensated for by increases in rainfall during the winter. The National Rivers Authority has resurrected plans for long distance trans-fers of water from northern and western areas to the south-east.

Future regional droughts as a result of climatic change may also lower the level of several major lakes and inland seas. The area of Lake Chad has dwindled alarmingly since the 1960s. The Aral Sea has lost 60 per cent of its volume in the last thirty years. It was the fourth largest inland sea in the world but its average inflow has been reduced from about 70 km³ a year to around 15. Much of the change has been due to the abstraction of water for irrigation but it highlights the sensitivity of such seas to changes in their water balance. The Caspian Sea may be threatened in a similar manner. On the other hand some countries face an increased risk of flooding due to more intensive monsoons, notably Bangladesh, but also parts of south-east Asia, India and East Africa.

IMPACTS ON AGRICULTURE

With continuing population growth, agriculture worldwide is already under increasing pressure without any changes in climate threatening it even further. Climatic variability becomes a greater and greater problem as farmers need to reach higher levels of output in order to continue in business. In South Australia, for example, a farmer needed to harvest 0.2 tonnes of wheat per hectare in the 1880s in order to survive. By the 1930s this had risen to 0.4 tonnes, by the 1960s 0.6 and by the 1980s 0.8 tonnes.

A warmer world, one where, on average, precipitation is expected to increase, has sometimes been presented as being beneficial to agriculture overall. In terms of agricultural productivity there may be winners as well as losers under new climatic conditions but the unreliability of the regional-scale scenarios produced by current GCMs makes it impossible to determine with any confidence which regions these are likely to be. More realistically, however, even with the benefits of carbon dioxide fertilization, agriculture in most areas of the world will face adjustments and, in some areas, considerable difficulty. Combinations of higher temperatures, shortage of moisture due to higher evaporation rates even if precipitation increases, and increased flood risk in some areas are probably the greatest threats to agriculture, with an increase in crop and animal pests also a possibility. However, the relationship between climatic change and its impact on agriculture is frequently a non-linear one. Small changes in temperature or moisture conditions may take agriculture across a critical threshold and have disproportionately large effects on production. The study of climatic impacts on agriculture is further complicated because different crops in the same region will respond differently to changes in climate, while the same crop in different areas will also show contrasting responses. For example the temperature rises suggested by GCM scenarios for a doubling of carbon dioxide would probably benefit the growth of barley in northern Finland due to higher temperatures but might reduce yields in southern Finland due to moisture stress (Parry et al., 1988).

Even in the tropics, where temperature rises are likely to be modest and precipitation is predicted to increase, higher loss of moisture from plants and soil may make things more difficult for farmers. Over the Earth as a whole plant growth is restricted by low rather than high temperatures so that warmer conditions might seem to be beneficial. However, in many tropical and temperate areas crop yields are limited by the availability of moisture rather than by temperature factors. There is a need to look at climatic impacts on agriculture at a regional scale, but the fact that changes in precipitation are difficult to predict complicates the exercise. Because detailed models of possible future regional climates are not available, agriculturalists often have to take an indirect approach by looking at current agricultural systems and deciding to which types of climatic change they are particularly vulnerable. It has often been suggested that the adaptability of farmers and developments in technology including new crop varieties will offset the effects of climatic change, in developed countries at least. Perhaps so, but there may be a serious danger of being too complacent and there may be some nasty surprises in store, especially regarding increases in crop pests.

From this, it will be clear that assessing the impact of climatic changes on agriculture is far from straightforward. Many climatic variables can influence crop growth and the relationship between climate and crops may operate indirectly. For example, climate may affect crops through the distribution of soil moisture, the impact of pests or by drier conditions encouraging soil erosion and reducing agricultural potential. Moreover, the interaction between climatic change and crop production cannot be fully appreciated without taking into account the wider frameworks of decision-making at the level of the farm, region and state. It is one thing to produce theoretical models which handle the physical links between climate and crops. It is much more difficult to build in the political, economic and social elements which combine to influence the responses of farmers and governments. Warmer conditions in southern Finland, for example, might increase yields of barley and oats simultaneously but the effect on total production of these cereals would depend very much on whether current levels of government support for farmers were continued (Parry et al., 1988). In some northern areas over-production under warmer conditions could force governments to introduce or extend 'set-aside' policies to take more land out of cultivation.

Although the climatic variables which can affect crops are numerous and complex, many studies have been based on considering the impact of changes of temperature and precipitation on crop yields associated with an effective doubling of carbon dioxide levels. Instead of quoting figures for possible temperature rises it is sometimes more meaningful to draw analogues. With a doubling of carbon dioxide levels many GCMs have suggested that Iceland could become as warm as Grampian, St Petersburg as warm as the Ukraine and Finland as warm as North

Germany. On this analogy yields of hay in Iceland might increase by 50 per cent in line with those current in north-east Scotland though this would probably be offset to some degree by a higher incidence of pests (Parry, 1990).

Theoretically the temperature increases which have been suggested for a doubling of carbon dioxide would increase the length of the growing season in high latitudes and shift the limits of crop production polewards. A 1°C increase in mean annual temperature represents a shift in cereal cropping 150–200 km north and 150–200 m upwards in the mid-latitudes of the northern hemisphere (Fig. 6.6). On this basis in North America the corn belt would move northwards by about 175 km for every 1°C rise in mean annual temperature while rice cultivation in Japan could move upslope several hundred metres. A longer growing season in middle and higher latitudes in the northern hemisphere might seem a bonus for farmers but there are hidden dangers. Early-maturing crops in such areas are more likely to suffer from heat stress so that a change from spring wheat to winter wheat might have to be made in the northern part of the US Great Plains and in Canada. Higher temperatures may encourage grain to mature faster resulting in a smaller proportion of the biomass of crops being stored in the seed, thus reducing yields. A 3°C rise in average annual temperatures in eastern England might in this way produce a 10 per cent drop in wheat yields. Additionally, many temperate crops such as oats and barley require a period of low temperature in winter to start or accelerate the flowering process.

Other factors would also intervene to reduce the benefits of temperature increase in higher northern latitudes. The poor quality of northern soils would limit the extent to which wheat cultivation could push northwards in Canada, Scandinavia or Russia. Soils in Siberia have greater potential for cultivation, with the northward shift of cultivation limits, than those of the Canadian shield. Warmer conditions would encourage crop pests and enable them to infest new areas. Iceland is currently too cold to suffer from potato blight but this might change in the future. The European corn borer, a pest which affects maize, could move northwards by between 165 and 500 km. with a 1°C increase in mean annual temperature. Warmer conditions could also allow pests to spread earlier in the growing season causing more crop damage. Many livestock diseases presently confined to the tropics might also be able to spread into temperate areas.

The warm summers of 1989–92 saw several species of insects and spiders which normally thrive in warmer climates spreading across Britain. Clouds of aphids from the Continent crossed the Channel while woodlouse spiders from the Mediterranean spread in the London area as did scorpions. Termites are moving northwards through Europe and locusts have started to affect southern Europe. These infestations could well be temporary if colder conditions return but, if warming continues,

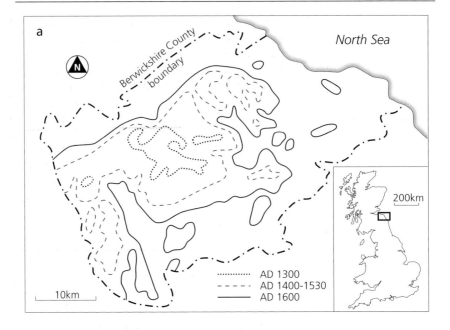

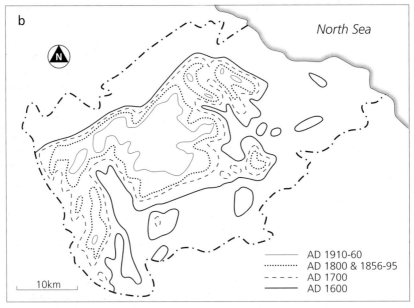

Fig. 6.6 Estimated changes in the climatic limits to crop production in part of south-east Scotland between AD 1300 and modern times, showing how small variations in mean temperatures can have large effects on cultivation limits in marginal areas. (After Parry, 1977)

they could be the advance guard of waves of insects which, because of their mobility, are quick to adapt to new climatic conditions.

Forecasts of higher yields with higher temperatures are based on the assumption that sufficient moisture will be available, but this might not be the case. In eastern England a 2°C temperature rise would cause a 9 per cent increase in evaporation if precipitation levels remained unchanged. This would greatly increase the soil moisture deficit in late summer and would probably necessitate extensive irrigation. Temperature rises in the US Midwest with a doubling of carbon dioxide levels might lead to a drastic decline in soil moisture which would seriously affect grain yields. The 1988 drought, which caused a 30 per cent drop in crop yields, affecting grain prices globally, may have been a foretaste of conditions which could become more frequent in future.

A study by the US Environmental Protection Agency has predicted that, with a doubling of carbon dioxide levels, a drop in corn yields of 50 per cent would occur in North Carolina, Tennessee, Georgia and Mississippi, 30 per cent in Ohio, Iowa, Nebraska and Kansas. This would be only partly offset by potential increases of up to 30 per cent in Texas, Oklahoma, Missouri, Michigan and Minnesota. Shortage of soil moisture would increase the demand for irrigation in the American Midwest but supplies of irrigation water are limited while increased abstraction for agriculture would have a serious effect on streamflow levels.

Most studies of the effects of climatic change on agriculture have been concerned with the impact on crop production. Much less has been done in trying to predict the impact of global warming on the carrying capacity of pasture and its effects on livestock. One study of Iceland suggested that the 4°C rise in temperature and a 15 per cent increase in precipitation predicted with a doubling of carbon dioxide would cause the growing season for grass to start as much as fifty days earlier and would improve the carrying capacity for sheep by around 2.5 times on areas of improved pasture and 50 per cent on rough pasture (Parry, 1990). As sheep farming is the main agricultural activity in Iceland improvements on this scale might lead to considerable savings by cutting the import of animal feed. Southern rangelands like Patagonia might enjoy similar benefits. In the case of the African grasslands the crucial variable is precipitation. Slight proportional increases or decreases could have considerable effects on livestock carrying capacity.

The adaptation of agriculture to climatic change involves three stages (Mount, 1994). First, the development of new crops, new rotations and different management practices to meet changing conditions. Second, the adaptation of farmers themselves to new circumstances and third the reaction of regional, national and international markets to changes in the supply of agricultural produce. How will farmers react to these predicted changes? Studies of the impact of climatic change on agriculture have usually concentrated on commercial rather than subsistence agriculture, on the physical impacts on particular crops rather than on the ability of farmers to perceive changes in the risks inherent in farming and adapt their strategies accordingly. Future temperature shifts may push the theoretical limits of the US corn belt north 175 km for every 1°C rise in temperature or make the cultivation of maize a commercial possibility in northern England, but such concepts are meaningless unless farmers respond by altering their patterns of cropping. A rise in average annual temperature, spread over several decades, may be of far less significance to farmers than the increased incidence of extreme conditions; heatwaves, droughts or a reduction in the occurrence of frost. The various scenarios relating to the response of crop yields to changing climatic conditions discussed in this chapter are based on the assumption that current technology and systems of management remain unchanged, but this is hardly likely. Suitable adjustments in farming systems could considerably reduce the impact of climatic changes but this would depend on governments altering their policies to encourage this. In the USA, for instance, current farm programmes often discourage adaptations such as switching crops or investing in water-conserving technology (Lewandrowski and Brazee, 1993).

Three main types of land use changes are likely to occur as a result of global warming; changes in the area under cultivation, in the types of crops grown and in cropping systems. Expansion of the cultivated area is likely in some northern areas of Europe, Canada, Russia and Japan, and more limited areas of the southern hemisphere including New Zealand and Argentina. Uphill shifts may also occur in some mountain areas. On the other hand, moisture deficits are likely to force a contraction of cultivation in many semi-arid regions such as the American Great Plains and the southern Mediterranean. Within the cultivated area higher temperatures are likely to cause a change to more heat-demanding crops, or those requiring less moisture where increased drought is a problem.

One crop which could undergo a dramatic shift northwards with higher temperatures is maize. It may become economically possible to cultivate maize far north of its present limits. Maize cultivation in north-west Europe is limited to areas with an Effective Temperature Sum of more than 850 degree days. Under present conditions, maize can be cultivated south of a line from the south-west tip of England to just south of Moscow. An equivalent doubling of carbon dioxide would shift this limit 200–350 km northwards, covering all the North European Plain including the whole of England, Ireland, southern Finland and southern Sweden (Fig. 6.7). But would precipitation be sufficient to allow an expansion of maize cultivation on this scale? A time-dependent scenario has suggested that the limits of viable maize cultivation may shift by 150 km per decade between the 1990s and 2030 and 240 km per decade between 2030 and 2060 (Parry, 1990).

In terms of changes in management there is likely to be increased pressure for irrigation to offset reduced moisture due to higher evapo-

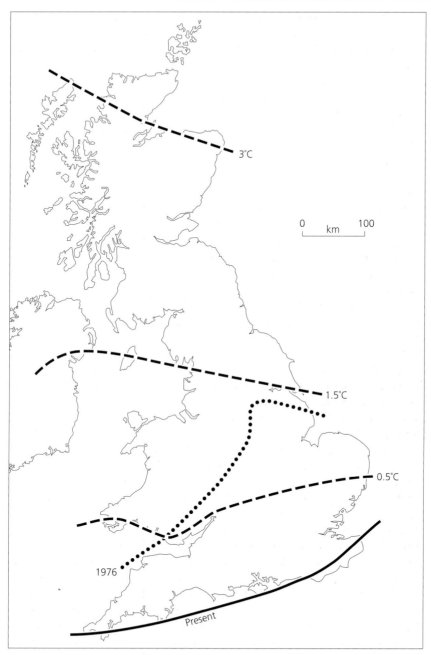

Fig. 6.7 Hypothetical limits to the ripening of grain maize in Britain with increases in mean annual temperatures. Present limits and those of the hot summer of 1976 are also shown. (After Parry, 1990)

transpiration. Providing irrigation water will increase costs and may not often be feasible in terms of available water supplies. While demand for more irrigation is likely to rise substantially in the American Great Plains it is also expected to become a more common feature of agriculture in western and southern Europe. Changes in agriculture will need to be linked to more careful management of hydrological resources. Overall output from the world's main grain-producing areas is likely to fall, especially in the USA but also in Canada and Russia. Within Europe, increased production of grain in areas like Britain and the Low Countries may be offset by falling returns in southern Europe.

In recent years a number of studies have examined the prospects for agriculture in different parts of the world. Most of these have focused on temperate areas and there have, as yet, been few regional studies of the possible responses of agriculture in tropical areas to climatic change. This has been partly due to a lack of reliable information at this scale from GCMs. Consideration of two case studies from the American Midwest and the Mediterranean will help to highlight the general problems which have already been considered in earlier chapters.

North America: the Midwest and the Great Plains

Concern has been expressed over the effects of climatic change on agriculture in the USA because of the central importance of US food surpluses in world trade. The drought of 1983 in the US corn belt reduced maize yields by 29 per cent compared with the previous year. This, combined with a reduction in the area of maize encouraged by the US government to limit over-production, led to an overall drop in maize output of 50 per cent. The worry is that drier and more variable conditions in this region in future would increase the scale of year-to-year fluctuations in food production (Parry et al., 1988). In most parts of the USA higher temperatures are calculated to lead to a reduction in soil moisture, an increase in heat stress and falling yields, except where crops are irrigated, apart from more northerly areas where temperature rather than precipitation is the main constraint on crop growth. In such areas yields of maize and soya beans could rise. On the Great Plains the predicted temperature rises linked to a doubling of carbon dioxide (between 3.8 and 6.3°C) and a corresponding 10 per cent fall in soil moisture might cut yields of maize by 15–25 per cent. A drop in yields of maize of up to 15 per cent in California has also been suggested.

It is not clear how much the carbon dioxide fertilization effect would help Great Plains farmers in offsetting a reduction in soil moisture for wheat cultivation. They are used to coping with periodic droughts but, if the frequency of dry conditions increases significantly, present farming systems may become non-viable. Higher temperatures and reductions in soil moisture would also increase the potential for wind erosion of soils. Canada, on the other hand, might be expected to benefit from possible temperature increases with an equivalent doubling of carbon dioxide due to an increase in the length of the growing season for spring wheat; ten days for each 1°C of temperature rise. There may also be a substantial shift in the winter wheat belt into Canada from the US Great Plains but, in the main areas of production, a fall in yields due to insufficient soil moisture has been forecast. In Sasketchewan it has been suggested that the growing season would be lengthened by between four and nine weeks and the period required for spring wheat to mature reduced by four to fourteen days (Parry et al., 1988). However, other aspects of climatic change would be less beneficial. Moisture reduction due to higher evapotranspiration rates would restrict crop growth during the summer and yields of spring wheat might decline by as much as 30 per cent with an increase of 3–4°C forecast with an equivalent doubling of carbon dioxide levels. This area produces some 18 per cent of the wheat traded in international markets so the problem is potentially serious. Winter wheat or maize with higher thermal requirements, potentially better able to resist spring and early summer drought, might expand at the expense of spring wheat but its yields have nevertheless been predicted to drop by 4 per cent.

In North America, as more generally, it seems unlikely that gains made by the poleward expansion of cultivation limits would be sufficient to balance losses due to a lack of soil moisture at lower latitudes. Heat stress on crops may also become a significant element in the American Midwest. Under present conditions crops suffer heat stress and yields fall off when temperatures exceed 35°C. A rise in average annual temperatures of under 2°C would increase the probability of five consecutive days with temperatures of over 35°C by a factor of three. Overall, for the USA, crop production would probably not be affected seriously enough to jeopardise domestic needs but there is a possibility that exports could decline markedly, with repercussions for the global trade in foodstuffs.

An integrated study of the MINK region (Missouri, Iowa, Nebraska and Kansas) of the American Midwest represents one of the most comprehensive and detailed investigations of the potential regional impacts of climatic change (Rosenberg, 1993). The current economic baseline of the region was first established. The use of GCMs was rejected because of their lack of detail at regional scales. Instead, the climate of the 1930s Dust Bowl era was used as an analogue for future conditions. During this period mean temperatures in the region were around 0.83°C higher than in the control period 1951–80. The analogue climate was then superimposed on the technological and economic conditions which applied to the period 1984–7. With regard to agriculture it was found that overall yields of corn and sorghum were reduced. Production of dryland wheat was unaffected and that of irrigated wheat rose with warmer conditions (Easterling et al., 1993). Farm-level adjustments using currently available low-cost technology eliminated 80 per cent of the negative impacts of climatic change on crop and livestock production. Impacts on forests in the area were more severe with a drop in forest biomass of 11 per cent due to moisture stress (Bowes and Sedjora, 1993). Water resources became much more scarce in the MINK region under the analogue with a drop in mean total streamflow to 69 per cent of that of the control period for the Missouri with effects on the availability of irrigation water, navigation, hydroelectric power generation, recreation and wildlife habitats (Frederick, 1993). Overall, however, the consequences of the analogue climatic change were modest; a decline of around 1–2 per cent in regional income and production (Bowes and Crosson, 1993). The research team admitted, however, that its study had examined the impact of climatic change on each element of the economy in isolation and then simply added up the figures. There was a suspicion that the combined sectoral impacts might be greater than the sum of their parts. There was also the possibility that actual climatic changes might be more severe than the analogue climate used (Crosson and Rosenberg, 1993). Limitations notwithstanding, this project points the way ahead for future regional impact studies.

The Mediterranean

For the Mediterranean, GCMs suggest possible temperature rises of up to 4°C with an equivalent doubling of carbon dioxide. Some models show a general increase in precipitation in the region, others a decrease. On a seasonal basis some models show a slight increase in winter precipitation but decreases in summer. Even where models suggest that overall precipitation might rise, the increase would be more than compensated for by greater evapotranspiration causing a soil moisture deficit possibly 15–50 per cent greater than today. Mediterranean agriculture already faces problems with a tendency for water deficits and the region is vulnerable to year-to-year variability in weather conditions. It is likely that, if a substantial temperature increase does occur, conditions in areas where winter rains are less reliable (such as North Africa and southern Spain) would spread northwards. Availability of water is the main factor limiting plant growth in the region today. The Mediterranean is vulnerable to aridity. Decreases in precipitation and higher evaporation will increase the areas where soil is affected by saline conditions, and the frequency of forest fires which already affect about 650,000 ha a year. Although conditions may become drier, more variable rainfall might produce severe periodic floods accelerating soil erosion. If warmer winters occur in the region more precipitation is likely to fall as rain rather than snow in the mountains around the basin leading to faster run-off and reduced water availability in late spring and early summer when it is needed for crop growth. Higher temperatures could allow a northward shift in the limits of cultivation of citrus fruits and olives, and a spread into the area of tropical crops such as avocado, banana, mango and sugar cane provided sufficient water was available.

Tourist numbers in the Mediterranean are predicted to double by about 2050 putting increased pressure on water resources and food supplies. There is already, with recent droughts, growing pressure on water supplies for irrigation. Of the available water in the Mediterranean, 72 per cent is used for irrigation against only 10 per cent for domestic use and 16 per cent for industry. Further erosion of deltas is likely to occur with rising sea levels and more intensive management of water resources within river catchments. The River Ebro, for instance, stopped expanding in about 1970 and is now steadily eroding due to water management schemes which have reduced sediment discharge to a tenth of what it was formerly (Jeftic et al., 1992). The delta is experiencing increasing saline penetration of its soils and higher sea levels may necessitate the pumping of water from low-lying land. Salinity increases in coastal areas may threaten the ecosystems of important wildlife areas such as the Carmargue.

Temperature rises may also alter patterns of pressure and wind frequencies affecting the circulation of water in the Mediterranean. This is already evident in the Adriatic as a result of the weakening in recent years of the Bora, the cold wind which blows off the Alps in winter. The head of the Adriatic is shallow with only a sluggish movement of water which depended on the Bora to set currents in motion. As this wind has weakened and almost failed in recent years, the sea has become stagnant with little mixing. During the late 1980s, beaches on the eastern coasts of Italy and in Yugoslavia were affected by the washing ashore of huge masses of partly decomposed algae. Rafts of algae covered hundreds of square miles of the Adriatic and produced a major slump in the tourist trade. Their occurrence was not a new phenomenon but the scale of recent algal blooms has been far greater than in the past. The algae have been absorbing so much oxygen from the waters of the Adriatic that fish stocks have been threatened. Nutrients from agricultural fertilizers washed into the sea by the River Po may have further encouraged the process of eutrophication.

The fifth consecutive year of drought in the Mediterranean was 1993 when wetland areas were drying out. Model scenarios suggest that such conditions could continue and intensify in future. Water shortages are already a problem, not just on the drier southern shores of the Mediterranean but also in France, Greece, Italy and Spain. Islands have even more severe problems; Malta already relies on desalinisation plants to produce water at three times the cost of pumping groundwater. Dams and irrigation systems have already caused a reduction in river discharges, lowering water tables and threatening wetland habitats such as the Coto Doñana on the Guadalquivir delta (Fig 6.8). In Spain there are plans for long-distance water supply grids to divert surpluses from the wetter north.

Fig. 6.8 Irrigated land in central Majorca. The increasing demand for irrigation water in the Mediterranean is already placing a severe strain on existing water resources. (Photo: I. Whyte)

WORLD FOOD SUPPLY

As well as having potentially important effects on agriculture at a regional scale, global warming also has major implications for world food supplies. These are already finely balanced due to population growth. Between 1954 and 1984 world grain output rose 2.6 times but this rise has levelled off since then. Only a few countries are net exporters of grain; the USA, Canada, France, Australia and Argentina are the main ones. In a good year world food production exceeds demand by around 10–20 per cent and surpluses are available for storage, but in bad years supply can fall to levels which are barely adequate to meet demand and stocks of corn can easily run out after bad years as happened in 1988 due to the American drought. In particular there is concern over the possibility of a substantial decline in yields of grain in a few key areas which supply much of the world's traded surpluses, notably the USA and Canada. Averaged over the globe it seems unlikely that total food production would be dramatically reduced as a result of forseeable climatic changes but alterations in the distribution of food production could nevertheless have serious economic, social and political consequences. Areas which already suffer from periodic food shortages could have their vulnerability dangerously increased, especially parts of Africa, south and south-east Asia and Central and South America. In these regions average incomes are among the lowest in the world and farmers' ability to adapt to changing conditions, especially by adopting new technology, is severely limited.

Deficits in precipitation and/or increased evapotranspiration could cause a substantial rise in the frequency of drought in many tropical and subtropical areas such as India where below-average rainfall in 1987 led to a drop in grain production from 152 to 134 million tonnes. If global warming lengthens the growing season in Canada and Siberia the US may lose some of its competitive advantage, creating domestic economic problems and also a shortage of exportable grain surpluses. The first year in which the US produced less grain than it consumed was 1981 – a warning of what warmer conditions might bring.

Price rises due to more uncertain patterns of production in key areas might reduce the ability of food-deficient countries to pay for imports and cut the amount of available food aid. The IPCC report agreed that food surpluses could be severely threatened by climatic change at regional if not global scales (Houghton et al., 1990). Predictions for the future are difficult to make, however, because of the problems of assessing the impact of new technology and allowing for the adaptability of farmers. These responses might well maintain agricultural production at existing levels – but at what cost? A recent study of the security of future global food supplies under changing climatic conditions was relatively optimistic, suggesting that, while global warming would tend to increase the already unequal distribution of food supplies (Rosenzweig and Parry, 1994), overall world production would not suffer too severely. However, this scenario depends on the full effects of carbon dioxide fertilization being attained which, it is clear, may not be the case. Nor do the authors consider possible limits on water supplies for irrigation. Without these benefits it has been suggested that world food production might fall by between 11 per cent and 20 per cent with an equivalent doubling of carbon dioxide (Pittock, 1994) In addition, the study used detailed crop growth models for specific locations and extrapolated them to cover entire regions. The authors' focus on temperate crops may have led them to overestimate the negative impacts of climatic change on tropical regions as many important tropical crops were not considered (Reilly, 1994).

This chapter has highlighted just some of the impacts which climatic change may have on regions which are already under pressure from human activities. Global warming is likely to increase the effects of changes already set in motion by poor environmental management. This emphasises the need to respond positively to the potential threats posed by global warming. The next chapter will consider what steps have already been taken in this direction and the potential for reducing greenhouse gas emissions as well as simply adapting to the changes which global warming may bring.

7

Responding to Global Warming

The IPCC report of 1990 emphasised that climatic change was a global issue requiring truly global responses and action. Such action would be likely to have considerable economic and social impacts. Furthermore, the IPCC stressed that both developed and developing countries had a shared responsibility to deal with the problem (Houghton et al., 1990).

If it is accepted that the threats posed by global warming are real, then two major approaches are possible. First, the scale of anthropogenic greenhouse gas forcing could be reduced by cutting emissions. Second, mankind could accept climatic changes as inevitable and attempt to adapt to them. Both approaches require as much knowledge as possible about the workings of the climate system and its interactions with the rest of the environment. Both approaches also need to acknowledge that climatic change can only be tackled by concerted global action, not least because the high cost of unilateral action could place individual countries at a severe economic disadvantage.

Given that temperature rises may be in the pipeline due to the effects of increases in greenhouse gas concentrations which have already occurred, it may seem hardly worth trying to reduce current and future greenhouse gas emissions. This pessimistic view has frequently been expressed by politicians, particularly in the light of the apparently high cost of reducing emissions on a scale sufficient to affect climate. Nevertheless, it should not be automatically assumed that attempts to cut emissions on a scale which would significantly reduce climatic change are unrealistic in technological and economic terms. While some warming of climate now appears inevitable due to past human actions, the amount and speed of future warming could be affected markedly by government policies relating to greenhouse gas emissions from energy production and use, agriculture and land use, as well as the production of CFCs (Fig. 7.1).

Analysis of the options for controlling greenhouse gas emissions is a rapidly developing field, though more progress has been made on

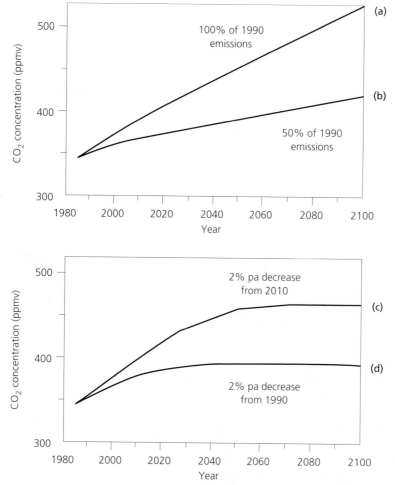

Fig. 7.1 Relationship between fossil fuel emissions of carbon dioxide and its atmospheric concentration where: (a) emissions continue to increase at 1990 levels (b) emissions are reduced by 50 per cent in 1990 and continue at this level (c) emissions are reduced by 2 per cent per annum from 1990 (d) emissions increase by 2 per cent per annum until 2010 then fall by 2 per cent per annum. (After Houghton et al., 1990)

reviewing the possibilities for reducing carbon dioxide and the CFCs than for other greenhouse gases. Complex economic concepts are involved in many of the calculations and space does not permit a comprehensive review of the various options which have been proposed, or the basis on which their costings have been calculated. Any attempts to tackle emissions must begin with political action so this chapter starts by looking at what has already been achieved.

THE POLITICAL FRAMEWORK

Political responses to scientific forecasts on greenhouse gas emissions and their likely effects have been cautious. As Schneider (1990) has pointed out, global warming is a good example of a problem in which the public need for reliable scientific information exceeds the current ability of science to provide it. A major difficulty is a widespread lack of understanding of how climate specialists work. Politicians often have a poor appreciation of how scientists approach this kind of problem and fail to understand their reluctance to make cast-iron, unqualified statements. There is an automatic tendency to equate climatic change with global warming, disregarding the fact that climatic changes on a range of time-scales are an essential part of the climate system. There is also a poor appreciation of the range of natural variability within the present climate.

While politicians are notorious for having short-term viewpoints and horizons bounded by the next election, (hence the acronym NIMTOO: Not In My Term Of Office), their caution is understandable, there is a danger of taking costly measures to meet contingencies which may never arise or, if they do occur, may not have the severe repercussions which were expected. Moreover there are strong vested interests in developed countries, especially relating to energy production, industry and transport, which are concerned about the adoption of policies which would increase their costs or which might affect expensive investments (Everest, 1988). In political terms the long lead times and high capital costs of, for example, new power stations, mean that it is impossible to make rapid changes of policy.

The interest of governments in climate is often in relation to short-term events such as the impact of floods, droughts and cold spells. Long-term climatic changes such as those we have discussed are not seen as being of comparable significance. Sir Crispin Tickell, UK ambassador to the UN, has highlighted this with the analogy of the 'boiled frog'. Suddenly drop a frog into boiling water and it will jump out if it can. Put the frog in cold water and gradually heat it up and you get frog soup! Human reactions to environmental problems are similar in that we respond positively to short-term threats but find the concept of long-term dangers much harder to grasp.

The general public, like the politicians it elects, finds it hard to grasp the nature of the problem due to the way in which it is usually presented. The huge scale of the potential threat is hard for people to appreciate unless they can relate it to their everyday lives. Equally, unless the impacts of climatic change are presented at this scale, people feel unable to do anything about them (Kearney, 1994).

Politicians perceive scientific opinion as divided, and many of them consider that scientists have overstated their case and exaggerated the threats posed by climatic change (Schneider, 1990). They point to the fact that temperature changes over the last century have been less than climatic models have predicted for the increase in greenhouse gases that has occurred. They also stress the high cost of taking counter measures against possible effects of greenhouse warming such as sea level rise. On the other hand, scientists like Stephen Schneider claim that, if we wait until the existence of global warming is proved conclusively, perhaps some decades in the future, by that time it will be much more difficult and expensive to do anything about it because of the greater amount of committed warming already under way.

Politicians tend to consider that, with the amount of uncertainty involved, any action now would be premature. If, as many people believe, the scale of future global warming and its impacts have been overestimated and if counter measures are adopted which are 'over the top' relative to the changes which do occur then a lot of money and effort will have been wasted. This cautious 'wait and see' approach is also a feature of many other environmental planners, such as water resource managers (Wyman, 1991). If, however, global warming does occur more or less at the rate predicted then adapting to changes might be more sensible and cost-effective than expensive counter measures. The attitude of the US government in recent years has been that it would be cheaper to adapt to climate change rather than restructure the whole energy system. This view has also been echoed in the US by the business community. Adapting to climatic change would, however, be easier for the developed world than for developing nations.

This view is a counter to the 'no regrets' policy advocated by many scientists who believe that lack of complete certainty must not be used as an excuse for inaction and that even a small chance of rapid climatic change warrants taking precautionary measures. Under such a policy heavy short-term costs are accepted in order to reduce emissions of greenhouse gases substantially in the hope of uncertain long-term benefits on the basis that, by the time watertight proof of the occurrence of global warming becomes available, climatic change may have gone too far for remedial measures to be applied. Even if global warming turns out to be less drastic than forecast such an approach would still have considerable environmental benefits.

Beyond rhetoric?

Lady Thatcher may have called global warming 'a serious threat to the world' and US Vice-President Al Gore recognised it as 'the most serious crisis we have ever faced' but has this rhetoric actually achieved anything? Some progress has indeed been made. The Montreal Protocol of 1987 and the follow-up London Agreement of December 1990 represent the first international efforts to reduce emissions of greenhouse gases, though this was achieved because of concern over ozone depletion rather than global warming. Nevertheless these agreements are important in demonstrating that international co-operation is possible once the seriousness of the problem has become apparent. However, global warming presents a much more difficult problem for political action. In the case of CFCs, production and end uses were relatively limited and clearly identifiable. Substitutes for CFCs already existed and were rapidly deployed as replacements. The other major greenhouse gases, such as carbon dioxide and methane, are produced by a diversity of activities by a far greater number of people. Measuring carbon dioxide and methane emissions from various sources or from different countries is also much harder than for CFCs.

Apart from CFCs, most of the efforts undertaken so far have been directed towards trying to reduce emissions of carbon dioxide, the most important man-enhanced greenhouse gas and one whose effects are long lived (Chapter 3). On the other hand, because of its short life, cutting methane emissions by 10 per cent would stabilise atmospheric concentrations at present levels much more rapidly. The degree to which opinions regarding policy decisions on greenhouse gas emissions have changed within a few years can be appreciated by comparing a 1988 paper on the greenhouse effect and issues for policy-makers prepared for the Policy Studies Institute (Everest, 1988) with the 1994 government response to the Framework Convention on Climate Change (HMSO, 1994). The main conclusion of the first paper was that governments should take no action and wait for the results of further research. It also suggested that emissions from greenhouse gases such as methane and nitrous oxide were not amenable to control, and dismissed out of hand suggestions for developing afforestation and renewable energy sources as ways of slowing down the increase of carbon dioxide. The 1994 statement, by contrast, has a whole series of targets for emissions reductions embracing all major greenhouse gases as well as the enhancement of greenhouse gas sinks.

An important milestone in political action towards tackling the problems of climatic change was the 'Earth Summit' at Rio in 1992. One element of this was the signing of a Framework Convention on Climate Change. The Convention, despite its limitations as discussed below, represents the most ambitious effort ever to control human activities in the interests of protecting the environment, the start of serious worldwide attempts to address the problems of climatic change. The aim of this was to stabilise greenhouse gas emissions at levels which would prevent dangerous climatic changes, and to do this sufficiently quickly for agricultural production and ecosystems not to be threatened, and economic development not to be disrupted. The Convention was considerably watered down before some countries, notably the USA, would sign it. This involved dropping legally binding commitments to specific targets for emissions reduction. However, although the response was less positive than originally hoped for, the

Convention has led to more detailed meetings which have begun to fill in the general structure established at Rio.

The developed countries agreed to the aim of reducing emissions of carbon dioxide and other greenhouse gases not covered by the Montreal Protocol to 1990 levels by AD2000. It is important to appreciate that this is not a legally binding commitment. Developing countries did not make any such agreements but they did undertake to make surveys of their greenhouse gas emissions and sinks and to produce individual response strategies to climatic change. The Convention accepted that, to allow economic development, energy consumption in developing countries would have to grow, producing rises in greenhouse gas emissions. Effectively more severe emissions cuts by the richer countries were to be required to offset increases from poorer ones (Fig. 7.2). The refusal of developing countries to commit themselves to emissions reductions until the developed world had demonstrated its own commitment to action seriously weakened the agreement.

The Convention had been signed by 166 nations when it was closed for signature in June 1993. Turkey, Saudi Arabia and several other Middle Eastern countries had refused to sign. Within six months of the Convention coming into force developed countries were required to submit national plans detailing policies to reduce greenhouse gas emissions and

Fig. 7.2 Copper smelting plant at Chuquimata, northern Chile, emphasising the potential of industries in developing countries to contribute to the build-up of greenhouse gases. (Photo: Dr. R. Auty)

increase sinks. The Convention did not have specific provision for funding its implementation. Ultimately the participation of developing countries will depend on the support offered by the developed world. Financial resources were to be made available for developing countries to meet the costs of implementing measures to reduce greenhouse gas emissions. Extra funds were planned for countries facing particularly severe threats, such as those with low-lying coastal areas, semi-arid regions subject to drought and desertification, and others with fragile ecosystems such as forest and mountain areas. The financing of these measures was to be arranged through the World Bank's Global Environment Facility but the amount of funding was not clearly specified. The industrialised world was asked to contribute tens of billions of dollars a year in 'greenhouse aid' but so far little of this has materialised.

After Rio

What progress has been made since Rio? Inevitably there have been difficulties. In December 1993 the environment ministers of the EU met to try and reach agreement on the joint ratification of the Convention. It looks as if it is going to be increasingly difficult for the EC to meet the target of stabilising carbon dioxide emissions at 1990 levels by 2000. Joint ratification by all EU members was desirable so that some of the poorer members such as Portugal and Spain could continue to develop their industry and increase their carbon dioxide emissions. The aim was for this to be balanced by richer nations agreeing to cut their emissions to less than the 1990 level. Unfortunately this plan collapsed due to a lack of agreement on an appropriate formula for sharing the burden, especially as Spain insisted on increasing its carbon dioxide emissions by 25 per cent by 2000. Britain agreed unilaterally to reduce its emissions to 1990 levels by 2000 but has refused to make further cuts to compensate for increased emissions by poorer countries.

A range of targets for emissions reductions has already been agreed by various countries. The Netherlands plans a reduction in carbon dioxide emissions of 3–5 per cent by 2000, relative to 1990, Belgium 5 per cent, Austria and Denmark a 20 per cent cut. New Zealand has agreed on a 20 per cent cut on 1990 emissions levels by 2000. France and Japan have agreed to stabilisation on a per capita rather than a total basis by 2000. Most countries' targets only involve carbon dioxide but Australia, Canada and the USA include methane and nitrous oxide. Germany's target is a cut of up to 30 per cent by 2005. Improvements in the efficiency of energy use and the run-down of inefficient industries in the former East Germany since unification provide considerable scope for emissions cuts. The feasibility of such targets is indicated by the fact that carbon dioxide emissions since 1979 have declined by 18 per cent in The Netherlands, 22 per cent in Germany, 26 per cent in Denmark and 27 per cent in Belgium (Bach, 1994).

Moves within the EU to impose a tax on energy, phased in over seven years from 1994, to boost energy conservation are essential if the EU is to meet the goals agreed at the Rio summit. These were blocked by the UK in March 1993. With the demise of the Clinton administration's efforts to impose a new energy tax in the US, plans for an EU carbon tax have been shelved for the present. Norway, Sweden, and Finland, countries which are not in the EU, have already imposed carbon taxes, as have Denmark and The Netherlands. But in all cases, the taxes are small or industrial fuel use is exempt.

Plans for major new road building programmes and nearly 150 new power stations within the EU look set to increase rather than reduce total emissions regardless of policies of reduction which may be pursued by individual countries. In November 1994 the British environment secretary, John Gummer, rejected the recommendations of a Royal Commission on Environmental Pollution that the price of petrol should be doubled within ten years and spending on trunk roads and motorways cut to try and halt the growth of traffic.

The US has also failed to deliver its promises. A plan produced in 1993 by President Clinton as the American response to the Convention has been heavily criticised because it relies on voluntary measures rather than enforcing new standards for carbon dioxide emissions. In particular the president appears to have given way to the car and oil lobby by failing to bring in new fuel efficiency standards. This reflects the widespread view in the US that it will be cheaper to adapt to global warming than to restructure the entire energy industry. The proposals include improving energy efficiency in homes and for electrical appliances as part of a target of cutting emissions by 100 million tonnes a year. Proposals to cut carbon dioxide emissions by 2000 were also opposed by Japan on the grounds that, as they were already much more energy efficient than the US, it would cost them far more to make the same proportional reductions.

Britain's response to the Convention is a plan to return emissions by 2000 to 1990 levels by a reduction of 10 million tonnes of carbon a year. The government has preferred to encourage use of market forces rather than adopting direct regulation, such as in curbing vehicle emissions, but the problem is that demand in many of these areas is not elastic. The plan is to save 1.5 Mt C on the supply side through the development of renewable energy sources and combined heat and power, using waste heat from electricity generation which improves the fuel conversion ratio of such plants to 70–90 per cent against 35–50 per cent for conventional ones. Nuclear energy is not expected to make a significant contribution to emissions reductions as only one new power station, Sizewell B, is scheduled to open before 2000. Renewable energy sources, which currently provide 1 per cent of Britain's electricity, are expected to provide 3 per cent by 2000.

Savings should also be made by switching from coal to gas-fired power stations. By 2000 gas fired power stations should produce about a quarter of Britain's electricity (against none in 1990) while the contribution from coal should fall from 68 to 47 per cent over the same period. These savings are not included in the calculations. It is planned that savings of 8.5 Mt C will come from measures on the demand side, 4 Mt from energy conservation in the home, encouraged by the introduction of VAT on domestic fuel, and new efficiency standards for household appliances and buildings. Savings of 2.5 Mt should come from businesses and 1 Mt from the public sector. A further 2.5 Mt will come from transport with plans for duties on fuel to be increased by at least 5 per cent a year in future budgets. There will be some savings too due to the enlargement of carbon sinks. One million hectares of Britain's 2.3 million hectares of forest is less than forty years old and should accumulate about 2.5 million tonnes of carbon a year, equal to about 1.5 per cent of the UK's carbon dioxide emissions.

There are also plans to reduce methane emissions from landfills and agriculture and nitrous oxide from industrial sources, especially nylon manufacture. However, whether the targets set will be achieved by this predominantly *laissez faire* policy is open to doubt. The target of recycling half of all recyclable household waste by 2000 is an example. Will small increases in petrol prices really cut emissions from traffic? In addition, many areas of emissions reduction are fortuitous, such as the expected drop in methane emissions from livestock due to a projected fall in dairy cattle numbers of 15 per cent by 2000, and a reduction in nitrous oxide emissions from fertilizer due to the expansion of set-aside land as a result of EU policies. Many areas of emissions reductions capitalise on expected market trends as a means of replacing the need for positive action.

A meeting of the Climate Change Convention in Geneva in September 1994 failed to agree on new targets for reducing greenhouse gas emissions, the first step in deciding who will be obliged to do what under the terms of the treaty and for assessing how different countries should contribute to emissions reductions. This meeting was a prelude to one in Berlin in March 1995 which provided the first opportunity to change the targets of the Convention since Rio. This meeting will be able to consider the findings and recommendations of the IPCC's second full assessment of climatic change, due to be published in 1995, which is expected to broadly confirm previous findings (Hulme et al., 1994).

A problem with the Convention is that it does not yet specify targets beyond 2000 though there is general agreement that the initial target of stabilising carbon dioxide emissions at 1990 levels does not go far enough. At the Geneva meeting Danish representatives indicated that their country was prepared to meet a target of 20 per cent reductions in carbon dioxide emissions by 2005 provided that other countries joined them. It is already clear that effective implementation of the Convention is likely to be hampered by the lack of detailed information on greenhouse gas emissions and sinks at a national level. If the Convention had been restricted to controlling carbon dioxide emissions from fossil fuel combustion this would have been easier but the existing data on emissions of methane and nitrous oxide at national levels are very inadequate, and there are problems of deciding where to draw the boundary between natural and anthropogenic emissions.

Who is responsible?

Political agreements on reducing greenhouse gas emissions are hampered by tensions between developed and underdeveloped nations over how responsibility for global warming should be allocated. Much of the increase in carbon dioxide and some other greenhouse gases such as CFCs has come from the developed world (Table 7.1) but, as noted in Chapter 3, this situation is set to change as developing countries increase their consumption of fossil fuels and reduce the areas of their tropical rain forests. Less developed countries have seen proposals for emissions reductions as a way of constraining their development and maintaining the hegemony of the industrialised nations. Bangladesh, of all nations, is perhaps the most immediately and catastrophically threatened by rising sea levels resulting from global warming but, with one of the lowest fossil fuel consumption rates in the world, the contribution of its population to the problem is limited, as is their ability to deal with it unaided (Table 7.2).

It is possible to calculate the contribution to global warming on a country-by-country basis as a measure of accountability. Details for some of the world's richest and poorest countries are shown in Table 7.2 but the limitations of such data can be readily appreciated. As indicated in

Table 7.1 Estimated contribution to radiative forcing changes

Greenhouse gas	During 1980s		1765–1990	
	Third world	Developed world	Third world	Developed world
CO_2 (total)	41%	59%	19%	81%
CO_2 (fossil fuel)	21	56	9	45
CO_2 (deforestation)	20	3	10	36
CH_4	57	43	57	43
CFCs	10	90	10	90
N_2O	20	80	20	80

(After Mintzer, 1992)

Table 7.2 Carbon emissions for selected countries

Country	Total carbon emissions 1950–86 megatonnes	Annual carbon emissions 1986 tonnes per cap
USA	37,284	5.01
USSR	22,039	3.59
Germany (united)	9,123	3.50
China	8,448	0.53
Japan	6,924	2.11
UK	5,922	2.94
France	3,646	1.79
Canada	2,911	4.09
India	2,184	0.19
Pakistan	192	0.13
Bangladesh	44	0.03
Ethiopia	10	0.01
Uganda	7	0.01
Nepal	3	0.02

(After Hayes and Smith, 1993)

Chapter 3 there is much uncertainty about the size of many greenhouse gas sources, such as carbon dioxide emissions from deforestation. Equally, the size of many sinks is only known within a wide range of error. But should this relate to current emissions or include past ones as well? Should emissions be calculated on a per capita basis rather than in total for each country? A per capita calculation would greatly reduce the contribution of third world countries such as China and India. Should a country's gross or net emissions be considered, or its ability to absorb carbon dioxide through natural sinks offset against its emissions? Can a unit of methane from a Chinese paddy field be considered 'equal' to a unit of carbon dioxide emitted by a motor vehicle in a western city?

The debate over methane releases highlights the sensitivity of developing countries and the inadequacies of current emissions data. The US Environmental Protection Agency recently estimated that rice cultivation in India was responsible for the emission of 37.8 million tonnes of methane a year, making India the world's leading methane generator. However, Indian scientists, undertaking a survey of their own, suggested an emissions rate of only 4.3 million tonnes. The discrepancy, they claimed, was due to the previous high figures being based on limited numbers of measurements of methane release from small plots of rice in other countries. Of India's rice fields, 60 per cent are irrigated lands or dry uplands rather than waterlogged paddies which produce twenty times as much methane (Minami and Meve, 1994). This demonstrates the shaky foundations on which many emissions estimates are built. To meet the needs of growing population, world rice production will need to rise by

Table 7.3 Apportionment of carbon dioxide emissions reductions

Industrialised countries	
Strong economies	47%
Less strong (eg: Eire)	2
Weak (E Europe)	25
Middle eastern oil producers	2
Countries in transition (eg: S Africa, S Korea)	4
Developing nations	20
	100

(After Bach, 1994)

around 65 per cent over the next thirty years, producing more methane unless careful management schemes are widely introduced.

National figures for emissions of all main greenhouse gases have now been produced (for example, Subak et al., 1993) but the authors admit that they involve major assumptions and uncertainties. Various ways of allocating the burden of carbon dioxide reductions have been put forward. A recent one (Bach, 1994) is shown in Table 7.3. However, when some less strong EU members such as Spain want a 25 per cent increase in carbon dioxide emissions to be acceptable the feasibility of such an apportionment, in the short term at least, is questionable.

The view, which has often been expressed by leaders of third world countries, that the developed nations should start to tackle the problems of emissions reductions themselves before trying to dictate what the developed world should do, has much to commend it in moral as well as economic terms. Stopping tropical deforestation is often urged as a priority in relating to reducing carbon dioxide emissions but, as noted in Chapter 6, compared with releases from fossil fuels this is strictly a sideshow as far as global warming is concerned, however desirable it may be on other grounds.

An aspect of the Convention which has attracted a lot of attention is the possibilities it creates for **joint implementation**. This involves developed countries working to reduce global emissions by investing in emissions reductions projects in developing countries. Joint implementation has been seen as a possible means of spreading emissions reduction technology but there is the danger that it may be misused as a way of avoiding significant reductions in developed countries by means of a kind of 'environmental colonialism'. Joint implementation is already being explored by the USA, Norway and Germany. It would be a suitable option for countries like Sweden where most energy comes from hydroelectric and nuclear power and where there is little scope for emissions reduction, or Japan which is a much more efficient user of energy than its major competitors.

The meeting in Berlin in 1995 considered the criteria for joint implementation in more detail. The case for it seems a reasonable one in
economic terms. If the amount of carbon dioxide emissions reduced in
China by a £1 million investment is five times the reduction that can be
achieved in western Europe for a comparable investment it might seem to
be a misuse of funds to apply them to Europe. Joint implementation offers
maximum returns for a given level of investment or the lowest cost way
of meeting specific emissions targets. However, there may be unforeseen
problems behind such schemes. The political instability of some developing countries could jeopardise joint implementation programmes while
some schemes, like afforestation, might be vulnerable to natural hazards
such as fire. There is clearly a place for suitable schemes of this type but
joint implementation cannot provide a solution to emissions reduction on
its own (Loske and Oberthur, 1994).

REDUCING GREENHOUSE GAS EMISSIONS

Chapter 4 considered some possible emissions scenarios for the future
based on various levels of emissions control. Another use for computer
models is to run them with different scenarios for the build-up of greenhouse gases to see how different policies on emissions reduction would
affect the rate of global warming and the time-scale over which it
occurred. The IPCC report suggested that, if emissions continue to increase
unchecked, the equivalent doubling of carbon dioxide concentrations
could be reached as early as 2025. With their low emissions scenario this
point would be reached by around 2040. To stabilise the greenhouse gas
content of the atmosphere would require a 15 per cent cut in methane,
60 per cent in carbon dioxide and 75 per cent in nitrous oxide. The 1988
Toronto conference on the Changing Atmosphere recommended a reduction of carbon dioxide emissions to 80 per cent of the level of the 1980s
by 2005. Building this into a model would delay the doubling of equivalent carbon dioxide concentration late into the next century – a rise by
2050 of only 15 per cent against 37 per cent in the IPCC BaU scenario.
This would still produce an estimated temperature rise of 2.1°C ±0.8°C
but this would nevertheless be a reduction of 0.5°C ±0.2°C.

Every time we drive a car, switch on a light or heat our homes we are
making a contribution to rising greenhouse gas levels and, in doing so, we
may be narrowing, ever so slightly, the options for future generations. A
reduction in emissions could be achieved by various means. One important step in slowing global warming would be controlling the size of world
population but space does not permit a review of this debate (Ehrlich and
Ehrlich, 1990). The potential already exists to make substantial cuts in
emissions, especially in the energy sector. Industry has already made an

important contribution by starting to phase out CFC production. Measures to reduce emissions would undoubtedly be expensive but many of them would have important and more tangible benefits in addition to reducing the speed and scale of global warming.

Fossil fuels

Not surprisingly there has been considerable opposition by less developed countries to ideas for controls on the burning of fossil fuels. The development of fossil fuel resources is seen by their governments as the only affordable route to economic development. Of greenhouse gas increases so far, 85 per cent have come from the developed world but use of energy by the developing world is increasing rapidly (Fig. 7.3). Fossil fuel energy saving measures in these countries is an important priority. Although politically difficult, reducing future emissions from less developed countries is one of the ways in which significant reductions in emissions could be made. Average emissions of carbon dioxide from developed countries have stabilised in recent years but those of developing countries have been

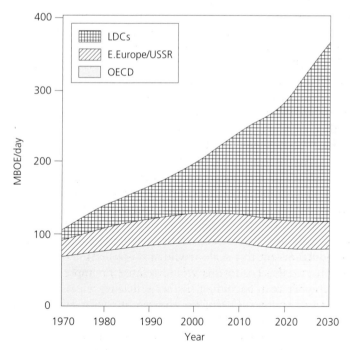

Fig. 7.3 Expected rates of growth in energy use (million barrels of oil-equivalent) in the OECD, eastern Europe and USSR, and less developed countries. (After Mintzer, 1992)

increasing by 4 per cent per annum, a figure which may soon rise to 5–6 per cent. The total primary energy demands of developing countries trebled between 1970 and 1990. They seem likely to double by 2010 and double again by 2025 (Mintzer, 1992).

Methane emissions

It has been suggested that better management of rice paddies could cut methane emissions by 10–30 per cent and that even greater reductions in emissions from livestock could be achieved. Improved agricultural practices might also reduce the release of nitrous oxide from fertilizers and carbon dioxide from soil erosion. Seepage of methane from landfills, coal mines and gas and oil production could also be reduced. It would be difficult, however, to make significant reductions in greenhouse gas releases for many activities, such as paddy rice cultivation and cattle rearing in the developing world. Much of the discussion which has appeared so far has been concerned with reducing emissions of the principal greenhouse gas affected by human activity, carbon dioxide. Here the potential for emissions reductions in areas such as electricity generation and use is much more clear.

The main approaches are seen as, first, energy conservation; reducing the amount of energy needed for particular purposes such as space heating, transport or power generation. Second, there is fuel substitution; switching from coal to natural gas, which has a lower carbon content, or to nuclear power and renewable energy sources which produce no carbon dioxide. Third comes reducing energy consumption and, finally, the removal of carbon after combustion by scrubbing, tree planting or before combustion.

In technological terms the greatest potential for emissions reductions lies in the energy sector which currently accounts for about 57 per cent of the radiative forcing from man-made sources (Houghton et al., 1990). Removing government subsidies and controls on energy prices worldwide might lower carbon dioxide emissions by as much as 20 per cent by 2050 (Fish and South, 1994). Using the most energy-efficient technology currently available, a 20 per cent cut in emissions by 2020 would be possible, but reductions on this scale would slow rather than halt the build-up of greenhouse gases. To do this would require far more drastic measures at a much higher cost, including further efficiency measures and a switch to different fuel sources. Of course, the use of energy-efficient technology has other benefits as well as cutting carbon dioxide emissions these include: reducing imports of oil, avoiding the development of nuclear power programmes with the attendant difficulties of waste disposal, reducing acid rain and air pollution by reducing the output of sulphur dioxide, carbon monoxide and nitrogen oxides as well as carbon dioxide.

In developed countries, 25 per cent of all carbon dioxide emissions are generated by energy used domestically. Electricity use in less developed countries is growing by 6–9 per cent per annum against 2–3 per cent in the developed world. At this rate of growth the generating capacity of the developing world will equal that of the developed nations by 2020. There is considerable potential for energy saving in domestic housing; an estimated 50 per cent in older homes, 25 per cent in newer ones, with even greater reductions possible in commercial premises and industrial processes in general. Recent studies suggest that there are good prospects for improving energy efficiency in lighting, internal combustion engines and building insulation which are likely to spread because they are financially attractive under present economic conditions. Cavity wall and loft insulation can each produce savings equal to 750–850 kg of carbon dioxide per household per year. There is also scope for improved fuel efficiency in vehicles though a considerable amount has been achieved since the oil price rises of the 1970s.

The trouble with transport

In the developed world the use of private cars and transport demands in general are likely to increase steadily into the next century. As GNP rises in many underdeveloped countries vehicle density will increase at an even greater rate. The relationship between vehicle density and GNP per capita is very close (Mintzer, 1992). As the demand for personal mobility grows there is also a tendency for freight traffic to switch from railways to roads. Transport problems in third world countries are concentrated in major cities. In some cities, like Bombay, private car ownership per capita has virtually doubled within a decade reflecting the problem that the development of urban mass transport systems have not kept pace with rising incomes. Improved technology will help slow the growth of greenhouse gas emissions from vehicles but a problem of this scale requires government policies not only to encourage more efficient technologies but also to control the growth of transport-related carbon dioxide emissions.

At present, transport policies, such as they are, in most developed countries are aimed at facilitating the unlimited use of private transport through the provision of infrastructure. In third world countries, however, this may actually aid fuel efficiency; a vehicle on a rough track may use 25 per cent more fuel than one on a metalled road. Governments could do a lot more to encourage the adoption of more fuel-efficient vehicles by taxation, either through annual road taxation fees or duties on fuel. This illustrates the need not only for more efficient internal combustion engines but for transport policies which may slow the rise in greenhouse gas emissions from vehicles.

As with other aspects of global warming, measures taken to reduce greenhouse gas emissions from transport will have other significant

Table 7.4 Car ownership for selected developing countries

Country	Car ownership per 1,000 persons	
	1985	projected 2025
China	0.6	20
India	2.1	16
South Korea	11.2	210
Brazil	63.0	245
Mexico	64.0	250

(After Mintzer, 1992)

environmental benefits such as the reduction of congestion and air pollution. Fuel efficiency improvements of up to 40 per cent may be possible within the next decade but road transport emissions of carbon dioxide are likely to go on rising. Even in a country like Britain with a high rate of car ownership traffic is expected to double between 1990 and 2025 (HMSO, 1994). Car ownership is expected to continue its spectacular rise in developed nations but the rate of increase is likely to be even more dramatic in many developing countries in south-east Asia and South America (Table 7.4). In developed countries, increased use of public transport, lower speed limits and charges for motorway driving would all make some contribution towards reducing the growth of emissions from transport, but major government investment and a change in attitudes towards the unlimited personal mobility offered by the private car would be required.

Industry

The IPCC report (Houghton et al., 1990) estimated that energy savings of up to 40 per cent could be made in many industrial processes. There is also scope for improving the efficiency of power generation as old power stations are phased out and replaced. The scope for improvement is demonstrated by the 40 per cent increase in energy efficiency achieved in the US economy in the seventeen years after the oil price rises of 1973. Further major savings in energy use could be made without sacrificing living standards. There is a need to make existing emissions-reducing and energy-saving technology more widely known and available. Energy-saving technology is developing fast but more government support for policies which support and encourage it is needed.

Substitution of fuels involves the short-term replacement of coal by natural gas, and the longer term goal of moving from fossil fuels to non-fossil fuels. At present, 77 per cent of the world's energy comes from fossil fuels, around 12 per cent from wood, crop waste and dung, 6 per cent from hydroelectric power and 5 per cent from nuclear power. Fossil fuels will continue to provide most of the growth in energy supply in the next twenty to thirty-five years. Coal provides half the electricity in developing countries and is likely to remain at least as important in the future. There is considerable scope in the long term for emissions reductions per unit output with a major switch to low-carbon or carbon-free electricity generation. Of fossil fuels, natural gas has a lower carbon content than oil and a substantially lower one than coal. Although natural gas has only about half the carbon content of coal its use releases methane which balances out some of the benefits of switching fuels. Moreover, resources of natural gas are relatively limited by present estimates. Industrialised countries are increasingly opting for natural gas-fired turbines with combined-cycle systems for new power stations. In the UK, twenty gas-fired power stations should be operating by 1996 using combined-cycle gas technology. Such stations are quicker and cheaper to build than coal-fired ones and require smaller sites. Advanced combustion techniques like these can reduce carbon dioxide emissions by up to 40 per cent compared with conventional coal-fired power stations. More efficient coal combustion technologies are also being developed which would be a useful way of reducing emissions from developing countries. Some potential exists for switching fuels and burning biomass waste and methane from landfills instead of fossil fuels. In terms of electricity generation, switching to renewable energy resources on a large scale is also a possibility.

Renewable energy

There is scope for developing renewable energy resources, including solar energy, wind power and hydroelectricity, to make a far more important contribution than they do at present. Indeed, to produce substantial emissions reductions renewable sources such as solar and wind power and biomass energy will need to be fully viable in commercial terms by the end of the decade and making a significant contribution to global energy generation by 2025 (Nilsson, 1992). Projections for the expected growth of the share of electricity produced by these means in underdeveloped countries are modest due to the present cost of the technology. However, wind energy is already competitive with energy derived from fossil fuels in many areas while improvements in technology continue to reduce costs. In California and Denmark, wind energy is already making a significant impact on energy generation.

Concern over the visual impact of wind farms in areas of high landscape value, such as the national parks of England and Wales, have so far limited their development in Britain where about twenty are currently in operation, including Europe's largest at Llandinam in Wales with 103 turbines on a 4 km^2 site generating 30 Mw of electricity – 300 sites of this size could supply 10 per cent of Britain's electricity and save the equivalent of 30 million tonnes of carbon dioxide a year. A recent discovery that the blades of turbines on hilltops can block or distort television signals may place further limitations on their siting. There have also been complaints about the levels of noise produced by wind turbines though new ones should be quieter than existing ones. Only modest price rises in fossil fuels would make a big difference to the competitiveness of renewable energy at present levels of technology. In Britain, the Department of the Environment has stated that 20 per cent of UK electricity could be wind-generated by 2025.

Despite the availability of low emissions technology and renewable energy sources, their uptake is currently reduced by structural problems. Overcoming the barriers which exist to the adoption of such technologies is a major prerequisite for their spread. Consumers may be unaware of the alternatives available to them due to lack of information. They may be unable to afford the capital cost of installing new equipment such as solar energy systems unless more effective financing packages are available. Unless energy prices rise substantially they may not have sufficient incentive to adopt new technology or energy-saving measures (Mintzer, 1992). Clearly these are areas where government policies are critical in influencing the uptake of new technology.

Hydroelectric power is very site-specific and in many developed countries most of the available locations have already been utilised. The era of building massive dams to generate hydroelectricity is already over in the USA. The construction of large schemes of this sort also has undesirable environmental consequences. HEP only provides 1 per cent of British electricity, mainly in Scotland. This amount could be doubled if planning permission was agreed. Tidal power generation is also feasible. In Britain the Severn, Solway and Mersey estuaries have the greatest potential and, in Northern Ireland, Carlingford and Strangford Loughs (Royle et al., 1994). Such schemes would, however, involve the destruction of valuable coastal ecosystems and wildlife habitats and would be opposed by environmental groups. The Severn scheme could generate up to 8 per cent of the electricity demands of England and Wales.

Few coastlines have better conditions for the generation of wave energy than those of Britain. In 1991 Britain's first wave energy unit was opened in Islay to demonstrate the feasibility of wave energy. A considerable amount of further development will be necessary to improve the energy capture process and make wave energy generation more economically viable (McIlwaine, 1994).

Scope for the large-scale development of nuclear power as a greenhouse gas-free source of electricity seems increasingly limited despite optimistic advertisements extolling the environmentally-friendly virtues of nuclear energy which have appeared in the British press in recent years. In global terms, the amount of energy generated by nuclear power continues to grow due to the opening of power stations which were first planned during the 1970s. Nuclear power provides a major share of the electricity used in some countries, including France and Japan, and its contribution to electricity generation in the underdeveloped world looks set to rise, if modestly, by the end of the century. However, high costs, including the decommissioning of obsolete power stations and worries about safety, have caused a drop in orders for new plants in many countries. Sweden is planning to phase out its nuclear power stations and further development of the German nuclear programme, which produces about a third of the country's electricity, seems unlikely due to strong political opposition. Nuclear power could make a substantial contribution to the reduction of greenhouse gas emissions but only with a new generation of safer reactors and more acceptable ways of disposing of radioactive waste.

Liquid fuel from biomass may be competitive with petroleum by the end of the century. Intensive biomass production for gas and liquid fuels using short-rotation techniques such as coppicing has been seen as a feasible alternative to energy derived from fossil fuels (Read, 1994). Although biomass generates carbon dioxide when used for energy production the next crop absorbs it back from the atmosphere. Coppicing trials in countries like New Zealand and the USA have shown that twenty dry tons of wood per hectare can be produced compared with four to ten tons for conventional forestry, and even higher growth rates may be possible. It has been found that 1 ha of intensive coppicing absorbs the carbon dioxide produced by burning 17.5 tons of coal or 15.5 tons of oil, after allowance is made for natural absorption. This would require an area of 500 million ha to counter carbon dioxide releases at 1990 levels, or an area only a little larger than the amount of land lost from tropical rain forest since the Second World War. The scale of this might sound too great to contemplate seriously. However, combined with the development of other renewable sources, biomass fuels appear to have considerable potential. Given that the EU is currently paying farmers to take land out of crop production there would probably be more land available for conversion to intensive coppicing, even in developed countries, than might at first be expected. A number of British farmers are interested in the possibility of planting extensive areas of willow to coppice for fuel and some schemes are already under way (Royle et al., 1994). However, the production of fuel from biomass is best suited to moist tropical environments and could provide a valuable boost to the economies of Third World countries. The unrestricted expansion of 'biomass farming' over land not required for agriculture could, however, have a severe impact on wildlife habitats. A difficulty with biomass fuel is that of engine redesign and the need for a supply infrastructure. This means that it is easier to replace fossil fuels used for space heating and industrial processes than in transport.

Many of these technologies are not quite economic at present but it would only take a minor shift in prices resulting from improved technology or a 'carbon tax' pricing policy to alter this situation. Over the next two decades the contribution of renewable sources to energy production is likely to increase markedly, but from a very low initial baseline. A major shift towards economies in which a significant amount of energy came from renewable sources or nuclear power would take twenty to forty years to achieve in many developed countries. The paths which individual countries follow will depend on their access to energy sources. In the USA the existence of large coal reserves and the desire for security of energy supplies has influenced government policy towards climatic change and emissions reductions. However, the large area of the USA may encourage efforts towards biomass, wind and solar energy generation. In western Europe there are substantial reserves of natural gas, making a large-scale shift from coal- to gas-fired power stations an attractive option. Japan, by contrast, has to import 80 per cent of its energy so than an emphasis on energy efficiency seems likely (Fish and South, 1994).

Deforestation and afforestation

Although there is a considerable amount of uncertainty regarding how much carbon dioxide is released annually from land use changes including deforestation, on an estimated release of about 1.6 GtC per year deforestation is responsible for only 13 per cent of greenhouse gas induced radiative forcing in the 1980s. Halting deforestation, even combined with large-scale afforestation, would only make a modest contribution to reducing future warming. Stopping deforestation by AD2000 would reduce the rise of carbon dioxide by 2050 by only 6 per cent. Nevertheless, combined with other measures, this would be worthwhile and would have more general ecological benefits. Prospects for reducing the rate of destruction of tropical rain forests and even a reversal of the trend allowing areas which have been less severely damaged to revert to some kind of forest are unclear. But what about creating more forests to balance or even outweigh losses in the tropics? A number of proposed solutions sound bizarre enough to be in the realms of science fiction. They may nevertheless be perfectly feasible technologically though their practicability and cost may be another matter.

Afforestation on a huge scale has been proposed as a way of increasing the size of the biosphere's carbon sink. One tree will absorb around one tonne of carbon dioxide over a twenty to thirty year period. Of course, afforestation will only lead to the absorption of carbon until the trees reach maturity, but if the wood was used as a source of energy replacing fossil fuel and replanting occurred, the effect would be long-term. The problem is that it would require the planting of an area the size of Australia to achieve anything useful. Each year a net 3–5 billion tonnes of carbon is added to the atmosphere. To store 1 billion tonnes of this in forests would require the planting of 1–2,000,000 km². The IPCC suggested that, if 10,000,000 ha of forest was planted each year, producing 4,000,000 km² by 2030, this would absorb around 1 Gt of carbon a year in some scenarios or in the region of 20 Gt by 2030, and about 80 Gt after 100 years, an amount equal to 5–10 per cent of emissions of carbon dioxide due to burning fossil fuels (Houghton et al., 1990).

Power stations

Carbon dioxide scrubbers for power stations are technologically feasible but very expensive. With present technology they use up to 4 per cent of the electricity produced. One estimate is that removal of 90 per cent of the carbon dioxide would double the cost of electricity. There are then problems of disposing of the carbon dioxide; storage in salt domes and the ocean deeps has been suggested (Wyman, 1991). Storage in disused aquifers could lock up carbon dioxide for a minimum of 10,000 years at a fairly low cost (Van Engelensburg and Blok, 1993). The environmental consequences of such an action are, however, unclear.

CFCs

In terms of estimated effects in reducing future temperature, the IPCC report considered that the banning of CFCs and more efficient use of energy were the two most effective options, producing reductions in future global warming of –0.52 and –0.45°C. Halting deforestation and afforestation programmes each contributed a reduction of only –0.2°C. Converting from oil and coal to natural gas reduced future rises by –0.28°C. The IPCC recommended a strategy of improving the efficiency of energy supply and consumption, and switching to fuels such as natural gas which reduced carbon dioxide emissions though they stressed that this was only a short-term expedient as natural gas cannot meet all energy needs and the continued large-scale use of coal, especially by developing countries is inevitable.

Clearly, an integrated approach to emissions reductions is needed; eliminating the production of CFCs, halting the destruction of rain forests and imposing increasing controls on carbon dioxide emissions in energy production would be a start. Such measures might stabilise the atmospheric greenhouse gas content by the middle of the next century but much more stringent controls on the release of carbon dioxide, methane and nitrous oxide would be needed to reverse the build-up of greenhouse gases. The reluctance of politicians to take direct action on reducing emissions is partly related to a belief that the introduction of such measures on a scale sufficient to be worthwhile would be extremely expensive. However, some recent studies (Read, 1994) have suggested that alternative energy-reducing technologies could be phased in and stabilise carbon dioxide levels with, on average, only a 10 per cent increase in energy prices over about 25 years – instead of the doubling of energy prices to achieve only a 20 per cent reduction in carbon dioxide levels by 2020 as suggested by some economists – without requiring major cuts in living standards in the developed world or blighting economic growth in developing nations. Whatever measures are taken by developed countries they will be insufficient on their own to deal with the problem of global warming without support from the developing world which needs to be helped to develop renewable energy technology.

♈ Conclusion

In an introductory survey of this kind it is impossible to discuss in depth all the themes which interlock and interact within a field as wide as the study of climatic change. Some topics, such as ozone depletion and defor-estation in the tropics, have been mentioned only in passing. Others, such as the potential effects of climatic change on human health and the distri-bution of disease, have had to be omitted entirely (Wyman, 1991). However, the complex nature of the climate system, along with its impact on other environmental systems and on mankind, have been emphasised and, in particular, the limitations of our current knowledge have been made clear.

It is evident that the question of whether global warming is actually occurring is not likely to be answered one way or another for many years. Even when an answer is forthcoming it is unlikely to be simple or straight-forward. Nor, until we have more detailed observations and more sophis-ticated ways of processing them, will we be able to predict future climates with any certainty. These problems make the study of climatic change at once frustrating and fascinating. It is also a field in which developments are rapid, which means that any state-of-the-art account will be rapidly overtaken by new discoveries. This book incorporated research findings which have appeared up to November 1994. By the time it is published a good deal of important new work is likely to have been undertaken. For example, recent work on the climatic impact of aerosols is likely to lead to major rethinking on the nature of future climatic changes. However, the basic concepts and problems involved change much more slowly and it is hoped that this book will provide the reader with a useful guide for a few years at least.

A consistent theme throughout this book has been that the effects of climatic change will be superimposed on, and are likely to exacerbate, other forms of human interference in the environment. Recent surveys of public opinion have demonstrated that a high proportion of respondents in the developed work have heard of global warming and the potential threat of the anthropogenically enhanced greenhouse effect. However, a far smaller percentage are able to identify carbon dioxide as the principal greenhouse gas whose concentration is increasing, or to link rising carbon dioxide levels with energy use (Jaeger et al., 1993). Haziness about the basic concepts involved has proved to be widespread even among people attending conferences on climatic change (Henderson-Sellers, 1990). Possibly because of the difficulties involved in grasping its scale and complexity, global warming is seen as a less important environmental threat by many people than freshwater and marine pollution or ozone depletion (Clayton, 1991). Unless the key issues involved in climatic change are more widely appreciated and understood in more depth, less positive action will be forthcoming, whether in terms of individuals adopting energy-saving technology in their own homes or governments taking concerted action to reduce emissions of greenhouse gases at an international level. It is the author's hope that this book may make some contribution to the wider understanding of the problems involved in climatic change.

Bibliography

General books on climatology

Barry, R.G. and Chorley, R.J. 1992: *Atmosphere, weather and climate* London: Routledge.

Henderson-Sellers, A. and Robinson, J.R., 1986: *Contemporary climatology* London: Longman.

General books on climatic change

Houghton, J.T., Jenkins, G.J., and Ephraums, J.J. 1990: *Climate change. The IPCC scientific assessment* Cambridge: Cambridge University Press.

Houghton, J.T., Callander, B.A., and Varney, S.K. 1992: *Climate change 1992. The supplementary report to the IPCC scientific assessment* Cambridge: Cambridge University Press.

Lamb, H.H. 1988: *Weather, climate and human affairs* London: Routledge.

Leggett, J. (ed.) 1990: *Global warming. The Greenpeace report* Oxford: Oxford University Press.

Maunder, W.J. 1992: *Dictionary of global climate change* London: UCL Press.

Nilsson, A. 1992: *Greenhouse Earth* Chichester: Wiley.

Schneider, S.H. 1990: *Global warming* Cambridge: Lutterworth Press.

Chapters 1 and 2

Bradley, R.S. and Jones, P.D. (eds.) 1992: *Climate since 1500* London: Routledge.

Dawson, A. 1992: *Ice age earth: late Quaternary geology and climate* London: Routledge.

Diaz, H. and Markgraf, V. (eds.) 1992: *El Niño: historical and palaeoclimatic aspects of the Southern Oscillation* Cambridge: Cambridge University Press.

Glantz, M.H., Katz, R.W., and Nicholls, N. 1991: *Teleconnections linking worldwide climate anomalies* Cambridge: Cambridge University Press.

Grove, J.M. 1988: *The Little Ice Age* London: Routledge.

Lamb, H.H. 1982: *Climate, history and the modern world* London: Methuen.

Lowe, J.J. and Walker, M.J.C. 1984: *Reconstructing Quaternary environments* London: Longman.

Philander, S.G. 1990: *El Niño, La Niña and the Southern Oscillation* London: Academic Press.

Roberts, N. 1989: *The Holocene. An environmental history* Oxford: Blackwell.

Wigley, T.M.L., Ingram, M.J. and Farmer, G. (eds.) 1981. *Climate and history* Cambridge: Cambridge University Press.

Chapter 3

Weubbles, D.J., and Edmonds, J. 1991: *Primer on greenhouse gases* Chelsea, Michigan: Lewis.

Chapter 4

Trenberth, K.E. 1992: *Climate system modelling* Cambridge: Cambridge University Press.

Chapter 5

Bird, E. 1993: *Submerging coasts.The effects of rising sea levels on coastal environments* Chichester: Wiley.

Doornkamp, J.C. (ed.) 1990: *The greenhouse effect and rising sea levels in the UK* Nottingham: M1 Press.

Tooley, M.J. and Jelgersma, S. 1992: *Impacts of sea level rise on European coastal lowlands* Oxford: Blackwell.

Warrick, R., Barrow, E.M. and Wigley, T.M. (eds.) 1993: *Climate and sea level change* Cambridge: Cambridge University Press.

Chapter 6

Bolin, B., Doos, B.R., Jaeger, J. and Warrick, R.A. (eds.) 1989: *The greenhouse effect, climatic change and ecosystems* Chichester: Wiley.

Parry, M.L. 1990: *Climate change and world agriculture* London: Earthscan.

Woodward, F. (ed.) 1992: *Global climatic changes: the ecological consequences* London: Academic Press.

Wyman, R.L. 1991: *Global climate change and life on Earth* London: Routledge.

Chapter 7

HMSO 1994: *Climate change: the UK programme* London.

Mintzer, I.M. (ed.) 1992: *Confronting climate change* Cambridge: Cambridge University Press.

Read, P. 1994: *Responding to global warming* London: Zed Books.

Glossary

ALBEDO The reflectivity of a surface, affecting the degree to which it absorbs or reflects solar radiation. Snow, ice and clouds have a high albedo, vegetation and the ocean surface a low albedo.

BUSINESS AS USUAL (BaU) SCENARIO One of the greenhouse gas emission scenarios developed by the IPCC (Houghton et al., 1990) which assumed that there would be few controls on emissions during the coming decades.

CFCs Chlorofluorocarbons.

COMMITTED WARMING Temperature rises which are still to come in the future due to changes in greenhouse gas concentrations which have already occurred, resulting from the time lag between rises in greenhouse gas levels and the corresponding amount of **EQUILIBRIUM WARMING**.

ENSO El Niño/Southern Oscillation (See Chapter 2).

EQUILIBRIUM WARMING The eventual rise of temperature which would result from a specific concentration of greenhouse gases at a particular time.

EQUIVALENT CO$_2$ CONCENTRATION The amount of carbon dioxide which would produce an amount of radiative forcing equivalent to that generated by all the different greenhouse gases acting together.

EUSTATIC Synchronous worldwide changes in mean sea level.

FEEDBACK A process by which the climate system reacts to changes acting upon it. These effects may amplify the original disturbance, destabilising climate and pushing it into a different mode (**POSITIVE FEEDBACK**), or damp down the original disturbance and stabilise the climate system (**NEGATIVE FEEDBACK**).

FORCING FACTORS Anything affecting the climate system which alters its radiative balance. For example, variations in solar radiation, orbital changes or the occurrence of major volcanic eruptions.

GCM General circulation model. Complex computer simulations of the atmosphere or the ocean used for simulating past, present or future climate.

GISP Greenland Ice Sheet Project.

GLOBAL WARMING POTENTIAL A measure of the relative radiative effect of equal emissions of different greenhouse gases, taking into consideration the length of time they stay in the atmosphere, and usually expressed relative to carbon dioxide.

GRIP Greenland Ice Core Project.

Gt Gigatonne: 1 Gt = 1 billion metric tonnes.

HOLOCENE The postglacial period spanning the last 10,000 or so years.

INTERSTADIAL Relatively short-lived warm periods during a glacial phase.

IPCC Intergovernmental Panel on Climate Change.

ISOTOPE A variant of a normal element whose atoms have a different mass because they have different numbers of neutrons. Isotope variations in sediments or ice can be used to reconstruct past climatic variations.

JOINT IMPLEMENTATION Projects by developed countries to reduce greenhouse gas emissions through investment in projects in developing countries.

PARAMETERIZATION In a **GENERAL CIRCULATION MODEL**, representing sub-grid scale phenomena such as cloud cover indirectly at grid square level using substitute data such as average humidity.

Pg Petagram: 1 Pg = 10^{15} grams.

PHOTOLYSIS Chemical decomposition under the action of light.

ppmv Parts per million volume.

ppbv Parts per billion volume.

pptv Parts per trillion volume.

PROXY DATA Indirect indicators of past climatic conditions such as tree rings or oxygen isotope variations.

REALISED WARMING The amount of warming which has occurred at any particular time between an increase in greenhouse gases and the attainment of **EQUILIBRIUM WARMING**.

SCENARIO A model representation of a climate which *could* occur under specified sets of conditions, not a prediction of what *will* occur.

SENSITIVITY The scale of response of climate to a perturbing influence such as greenhouse gas forcing.

STADIAL A relatively brief cold phase in which local ice advances could occur.

Tg Teragram: 1 Tg = one trillion grams or one million metric tonnes.

THRESHOLD A point beyond which small-scale changes in climate cause the climate system suddenly to change to a very different mode.

TIME-DEPENDENT MODEL Computer models in which increased greenhouse gas concentrations are phased in gradually over simulated time phases, involving the coupling of an atmospheric general circulation model to a dynamic model of ocean circulation.

TRANSIENT CLIMATE Climates due to greenhouse gas forcing which occur before an equilibrium climate is reached.

References

Allan, R.J. 1988: El Niño/Southern Oscillation influences in the Australasian region. *Progress in Physical Geography* 12, 313–48.

Alley, R.B., Meese, D.A., Shuman, C.C., Gow, A.J., Taylor, K.C., Grootes, P.M., White, J.W.C., Ram, M., Waddington, E.D., Mayewski, P.A., and Zielinski, G.A. 1993: Abrupt increase in Greenland snow cover at the end of the Younger Dryas event. *Nature* 362, 527–9.

Andrews, J.T., Davis, P.T. and Wright, C. 1976: Little Ice Age permanent snow cover in the eastern Canadian Arctic: extent mapped from LANDSAT satellite imagery. *Geografiska Annaler* 58A, 71–81.

Arnell, N.W., Jenkins, A. and George, D.G. 1994: *The implications of climate change for the National Rivers Authority* London: HMSO.

Bach, W. 1994: A climatic and environmental protection strategy: the road toward a sustainable future. *Climatic Change* 27, 147–60.

Ball, T.F. 1983: The migration of geese as an indicator of climatic change in the southern Hudson Bay region between 1715 and 1851. *Climatic Change* 5, 85–93.

Ball, T.F. 1992: Historical and instrumental evidence of climate: western Hudson Bay, Canada 1714–1850. In Bradley, R.S. and Jones, P.D. (eds.), *Climate since 1500* London: Routledge, 40–73.

Balling, R.C. et al. 1992: Climate change in Yellowstone National Park: is the drought-related risk of wildfires increasing? *Climatic Change* 22, 35–45.

Barber, K.E. 1985: Peat stratigraphy and climate. In: Tooley, M.J. and Sheail, G.M. (eds.), *The climate scene* 175–84.

Bergthorsson, P. 1985: The sensitivity of Icelandic agriculture to climatic variations. *Climatic Change* 7, 111–27.

Berk R.A., Schulman, D., McKeever, M. and Freeman, H.E. 1993: Measuring the impact of water conservation campaigns in California. *Climatic Change* 24, 233–48.

Bird, E. 1993: *Submerging coasts: the effects of rising sea levels on coastal environments* Chichester: Wiley.

Birks, H.J. and Birks H.H. 1980: *Quaternary palaeoecology* London: Arnold.

Birks, H.J. 1981: The use of pollen analysis in the reconstruction of past climates: a review. In Wigley, T.M.L., Ingram, M.J. and Farmer, G. (eds.). *Climate and history* Cambridge: Cambridge University Press. 111–38.

Blackford, J.J., Edwards, K.J. et al. 1992: Hekla-4 Icelandic volcanic ash and the mid-Holocene scots pine decline in northern Scotland. *The Holocene* 2, 260–5

Bluth, G.J.S., Schnetzlet, C.C., Frueger, A.J. and Walter, L.S. 1993: The contribution of explosive volcanism to global atmospheric SO_2 concentrations. *Nature* 366, 327–9.

Bolton, G.S. 1993: Two cores are better than one. *Nature* 366, 507–8.

Bonan, G.B., Pollard, D. and Thomson, S.L. 1992: Effects of boreal forest vegetation on global climate. *Nature* 359, 716–8.

Bond, G. 1993: Correlations between climatic records from North Atlantic sediments and Greenland ice. *Nature* 365, 143–7.

Bond, G. Heinrich, H., Broecker, W., Labeyrie, L., McManus, J., Huon, S., Tantschie, R., Clasen, S., Simet, C., Tedesco, K., Klas, M., Bonam, G. and Ivy, S. 1992: Evidence for massive discharges into the North Atlantic ocean during the last glacial period. *Nature* 360, 245–9.

Boorman, L.A., Goss-Custard, J.D. and McGrorty, S. 1989: *Climatic change, rising sea levels and the British coast* London: HMSO.

Bowes, M.D. and Sedjora, R.A. 1993: Impacts and responses to climatic change in forests in the MINK region. *Climatic Change* 24, 63–82.

Bowes, M.D. and Crosson, P.R. 1993: Consequences of climatic change for the MINK economy: impacts and responses. *Climatic Change* 24, 131–58.

Boyle, E. and Weaver, A. 1994: Conveying past climates. *Nature* 372, 41–2.

Bradley, R.S. 1985: *Quaternary palaeoclimatology* London: Allen and Unwin.

Bradley, R.S. and Jones, P.D. (eds.) 1992: *Climate since 1500* London: Routledge.

Brammer, H. 1993: Geographical complexities of detailed impact assessment for the Ganges-Brahmaputra-Meghna delta of Bangladesh. In Warrick, R., Barrow, E.M. and Wigley, T.M. (eds.), *Climate and sea level change* Cambridge: Cambridge University Press. 246–63.

Broadus, J.M. 1993: Possible impacts of, and adjustments to, sea level rise: the cases of Bangladesh and Egypt. In Warrick, R.A., Barrow, E.M., and Wigley, T.M.L. (eds.), *Climate and sea level change* Cambridge: Cambridge University Press. 263–75.

Broecker, W.S. 1994: An unstable superconveyor. *Nature* 367 414–5.

Budd, W.F., McInnes, B.J., Jenssel, D. and Smith, I.N. 1987: Modelling the response of the West Antarctic ice sheet to a climatic warming. In Van der Veen, C.J. and Oerlemans, J. (eds.), *Dynamics of the West Antarctic Ice Sheet* Dordrecht: Reidel 321–58.

Burgess, C. 1989: Volcanism, catastrophe and the global crisis of the late second millennium BC. *Current Archaeology* 117, 325–9.

Caldeira, K. and Casting, J.F. 1993: Insensitivity of global warming potentials to carbon dioxide emission scenarios. *Nature* 366, 251–2.

Campbell, I.D. and McAndrews, J.H. 1993: Forest disequilibrium caused by rapid Little Ice Age cooling. *Nature* 366, 336–8.

Cane, M.A., Eshel, G. and Buckland, R.W. 1994: Forecasting Zimbabwean maize yield using eastern equatorial Pacific sea surface temperature. *Nature* 370 204–5.

Cannell, M.G.R. and Hooper, M.D. 1991: *The greenhouse effect and terrestrial ecosystems in the UK* London: HMSO.

Catchpole, A.J.W. 1992: Hudson's Bay Company ships' logs as sources of sea ice data 1751-1870. In Bradley, R.S. and Jones, G.P. (eds.), *Climate Since 1500* London: Routledge, 17–39.

Catchpole, A.J.W. and Faurer, M-A. 1983: Summer sea ice severity in the Hudson Strait 1751–1870. *Climatic Change* 5, 115–39.

Chapellaz, J., Blunier, T., Raynauld, D., Barnola, J.M., Schwander, J. and Stauffer, B. 1993: Synchronous changes in atmospheric methane and Greenland climate between 40 and 8 kyr BP. *Nature* 364, 443–5.

Chester, D.K. 1988: Volcanoes and climate. *Progress in Physical Geography* 12, 1–35.

Clayton, K. 1991: Scaling environmental problems. *Geography* 76, 2–15.

Clayton, K. 1993: Adjustment to greenhouse gas induced sea level rise on the Norfolk coast. In Warrick, R., Barrow, E.M., and Wigley, T.M. (eds.), *Climate and sea level change* Cambridge: Cambridge University Press. 310–21.

Cohen, J. 1994: Snow cover and climate. *Weather* 49 no. 5, 150–5.

Cole, D.R. and Curtis-Monger, H. 1994: The influence of atmospheric CO_2 on the decline of C4 plants during the last deglaciation. *Nature* 368, 533–6.

Colls, K. 1993: Assessing the impact of weather and climate in Australia. *Climatic Change* 25, 225–45.

Coope, G.R. 1977: Quaternary coleoptera as aids in the interpretation of environmental history. In Shotton, F.W. (ed.), *British Quaternary studies: recent discoveries* Oxford: Oxford University Press, 55–68.

Crosson, P.R. and Rosenberg N.J. 1993: An overview of the MINK study. *Climatic Change* 24, 159–73.

Dacey, J.W.H., Drake, B.G. and Klug, M.L. 1994: Simulation of methane emission by carbon dioxide enrichment of marsh vegetation. *Nature* 370, 47–9.

Dawson, A. 1992: *Global climate change* Oxford: Oxford University Press.

Day, J.W., Conner, W.H., Costanza, R., Kemp, G.P. and Mendelssohn, I.A. 1993: Impacts of sea level rise on coastal systems with special emphasis on the Mississippi River deltaic plain. In Warrick, R.A., Barrow E.M., and Wigley, T.M.L. (eds.), *Climate and sea level change* Cambridge: Cambridge University Press. 276–96.

Den Elzen, M.G.J. and Rotmans, J. 1992: The socio-economic impact of sea-level rise on The Netherlands: a study of possible scenarios. *Climatic Change* 20, 169–95.

De Ronde, J.G. 1993: What will happen to The Netherlands if sea level rise accelerates? In Warrick, R.A., barrow, E.M. and Wigley, T.M.L. (eds.), *Climate and sea level change* Cambruidge: Cambridge University Press. 322–35.

Diaz, H. and Markgraf, V. (eds.) 1992: *El Niño: historical and palaeoclimatic aspects of the Southern Oscillation* Cambridge: Cambridge University Press.

Dury, G. 1984: Crop returns on the Winchester manors 1232-1340. *Transactions of the Institute of British Geographers* NS 9, 401–18.

Easterling, W.E., Crosson, P.R., Rosenberg, N.L., McKenney, M.S., Katz, L.A. and Lemon, E.M. 1993: Agricultural impacts of and responses to climatic change in the Missouri-Iowa-Nebraska-Kansas (MINK) region. *Climatic Change* 24, 23–61.

Ehrlich, P. and Ehrlich, A. 1990: *The population explosion* New York: Simon & Schuster.

Elkins, J.W., Thompson, T.M., Swanson, T.H., Butler, J.H., Hall, B.D., Cummings, S.O., Fisher, D.A. and Raffo, A.G. 1993: Decrease in the growth rates of atmospheric CFCs 11 and 12. *Nature* 364, 780–3.

Emiliani, C 1993. Milankovitch theory verified. *Nature* 364, 583–4.

Everest, D. 1988: *The greenhouse effect: issues for policy-makers* London: Policy Studies Institute.

Fairbanks, R.G. 1993: Flip-flop end to the last ice age. *Nature* 362, 495.

Fish, A.L. and South, D.W. 1994: Industrialised countries and greenhouse gas emissions. *International Environmental Affairs* 6, 14–44.

Fisher, M.J. et al. 1994: Carbon storage by introduced deep rooted grasses in the South American savannas. *Nature* 371, 326–8.

Flather, R.A. and Khander, H. 1993: The storm surge problem and possible effects of sea level changes on coastal flooding in the Bay of Bengal. In Warrick, R., Barrow, E.M. and Wigley, T.M. (eds.), *Climate and sea level change* Cambridge: Cambridge University Press. 229–45.

Flohn, H. 1985: A critical assessment of proxy data for climatic reconstruction. In Tooley, M.J. and Sheail, G.M. (eds.), *The climatic scene* 93–102.

Foley, J.A. et al. 1994: Feedbacks between climate and boreal forests during the Holocene period. *Nature* 371, 52–4.

Frederick, K.D. 1993: Climate change and impacts on water resources and possible responses in the MINK region. *Climatic Change* 24, 83–115.

Fritts, H.C. 1976: *Tree rings and climate* London: Academic Press.

Fritts, H.C., Lofgren, G.R. and Gordon, G.A. 1981: Reconstructing seasonal to century time scale variations in climate from tree ring evidence. In Wigley, T.M.L., Ingram, M.J., and Farmer, G. (eds.), *Climate and history* Cambridge: Cambridge University Press. 139–61.

Fritts, H. and Lough, J.M. 1985: An estimate of average annual temperature variations for North America 1602–1961. *Climatic Change* 7, 203–24

Glantz, M.H., Katz, R.W. and Nicholls, N. 1991: *Teleconnections linking worldwide climate anomalies* Cambridge: Cambridge University Press.

Gornitz, V. 1993: Mean sea level changes in the recent past. In Warrick, R., Barrow, E.M. and Wigley, T.M. (eds.), 1993: *Climate and sea level change* Cambridge: Cambridge University Press. 25–44

Grabherr, G. Gottfried, M. and Pauli, H. 1994: Climate effects on mountain plants. *Nature* 369, 448.

Grove, J.M. 1985: The timing of the Little Ice Age in Scandinavia. In Tooley, M.J. and Sheail, G.M. (eds.), *The climatic scene* 132–50.

Grove, J.M. 1988: *The Little Ice Age*. London: Routledge.

Grove, J.M. and Battagel, A. 1983: Tax records from western Norway as an index of Little Ice Age environment and economic deterioration. *Climatic Change* 5, 265–82.

Harding, A.F. (ed.) 1982: *Climatic change in later prehistory*. Edinburgh: Edinburgh University Press.

Hayes, P. and Smith, K. 1993: *The global greenhouse regime: who pays?* London: Earthscan.

Henderson-Sellers, A. 1990: Australian public perception of the greenhouse issue. *Climatic Change* 17, 69–96.

Henderson-Sellers, A. 1993: An Antipodean climate of uncertainty. *Climatic Change* 25, 203–24.

Henderson-Sellers, A. and Robinson, J.R. 1986: *Contemporary Climatology*. London: Longman.

HMSO. 1994: *Climatic change: the UK programme* London: HMSO.

Houghton, J.T., Callander, B.A. and Varney, S.K. (eds.) 1992: *Climate change 1992. The supplementary report to the IPCC scientific assessment* Cambridge: Cambridge University Press.

Houghton, J.T., Jenkins, G.J. and Ephraums, J.J. (eds.) 1990: *Climate change: the IPCC scientific assessment* Cambridge: Cambridge University Press.

Hughes, J. 1994: Catastrophes, phase shifts and large-scale degradation of a Caribbean coral reef. *Science* 265, 1547–51.

Hulme, M., Zhao, Z-O, and Jiang, J. 1994: Recent and future climate changes in East Asia. *International Journal of Climatology* 14, 637–58.

Idso, S.B. 1992: Scrubland expansion in the American south-west. *Climatic Change* 22, 85–6.

Ingram, M.J., Underhill, D.J. and Farmer, G. 1981: The use of documentary sources for the study of past climates. In Wigley, T.M.L., Ingram, M.J. and Farmer, G. (eds.), *Climate and history* Cambridge: Cambridge University Press. 180–213.

Jaeger, C., Durrenberger, G. Kastenholz, H. and Truffer, B. 1993: Determinants of environmental action with regard to climatic change. *Climatic Change* 23, 193–211.

Jeftic, L, Milliman, J.D. and Sestini, G. 1992: *Climate and the Mediterranean* London: Edward Arnold.

John, B.S. 1979: *The winters of the world* Newton Abbot: David and Charles.

Jones, P.D., Kelly, D.M. and Goodess, C.M. 1989: The effect of urban warming on the northern hemisphere temperature average. *Journal of Climate* 2, 285–90.

Jones, P.D., Wigley, T.M. and Wright, P.B. 1986. Global temperature variations 1861–1984. *Nature* 322, 430–4.

Jousel, J., Barkov, N.I., Barnola, J.M., Bender, M., Chapellaz, J., Genthon, C. Kotlyakov, V.M., Lipenkov, V., Lorius, C., Petit, J.R., Raynauld, D., Paisbeck, G., Ritz, C., Sowers, T., Stievenard, M., Yiou, F. and Yiou, P. 1993: Extending the Vostok ice core record of palaeoclimate to the penultimate glacial period. *Nature* 364, 407–11.

Karlen, W. 1973: Holocene glaciers and climatic variations, Kebnekaise Mountains, Swedish Lappland. *Geografiska Annaler* 55A, 29–63.

Kearney, A.P. 1994: Understanding global change: a cognitive perspective on communicating through stories. *Climatic Change* 27, 419–41

Keller, M., Veldkamp, E., Weitz, A.M. and Reiners, W.A. 1993: The effect of pasture age on soil trace gas emissions from a deforested area of Costa Rica. *Nature* 365, 244–6.

Kellogg, W.W. 1978: Global influence of man on the climate. In Gribbin, J. (ed.), *Climatic Change* Cambridge: Cambridge University Press 205–27.

Keppenne, C.L. and Ghil, M. 1992: Extreme weather events. *Nature* 358, 547.

Kershaw, I. 1973: The great famine and agrarian crisis in England 1315–22. *Past and Present* 59, 31–50.

King, J.C. 1994: Recent climatic variability in the vicinity of the Antarctic peninsula. *International Journal of Climatology* 14, 357–69.

Kingston, J.A. 1980-81: Daily weather maps from 1781. *Climatic Change* 3, 7–36.

Knox, T.C. 1993: Large increases in flood magnitude in response to modest changes in climate. *Nature* 361, 430–2.

Kuhn, M. 1993. Possible contributions to sea level change from small glaciers. In Warrick, R.A., Barrow, E.M., and Wigley, T.M.L. (eds.), *Climate and sea level change* Cambridge. Cambridge University Press. 134–43.

Kullman, L. 1994: Climate and environmental change at high latitudes. *Progress in Physical Geography* 18, 124–35.

Lamb, H.H. 1977: *Climate: present, past and future. Vol 2. Climatic history and the future* London: Methuen.

Lamb, H.H. 1982: *Climate, history and the modern world* London: Methuen.

Lehman, S. 1993: Ice sheets, wayward winds and sea change. *Nature* 365, 108–10.

Le Roy Ladurie, E. 1971. *Times of feast, times of famine* London: Doubleday.

Le Roy Ladurie, E. 1980: Grape harvests from the fifteenth to the nineteenth centuries. *Journal of Interdisciplinary History* 10, 839–49

Levenberger, M. and Siegenthaler, U. 1992: Ice age atmospheric composition of nitrous oxide from an Antarctic ice core. *Nature* 360, 449–51.

Lewandrowski, J.K. and Brazee, R.J. 1993: Farm programs and climate change. *Climatic Change* 23, 1–20.

Liu, H.S. 1992: Frequency variations of the Earth's obliquity and the 100 kyr ice age cycles. *Nature* 338, 397–9.

Loske, R. and Oberthur, S. 1994: Joint implementation under the climate change convention. *International Environmental Affairs* 6, 45–58.

Lowe, J.J. 1993: Setting the scene: an overview of climatic change. In Smout, T.C. (ed.) *Scotland Since Prehistory*. Aberdeen: Scottish Cultural Press, 1–16.

Lowe, J.J. and Walker, M.J.C. 1984: *Reconstructing Quaternary environments* London: Longman.

MacAyeal, D.R. 1992: Irregular oscillations of the West Antarctic ice sheet. *Nature* 359, 29–32.

Mackenzie, D. 1993: Where has all the carbon gone? *New Scientist* 141 1907, 30–33.

McGhee, R. 1981: Archaeological evidence for climatic change during the last 5,000 years. In Wigley, T.M.L. Ingram, M.J. and Farmer, G. (eds.), *Climate and history* Cambridge: Cambridge University Press. 162–80.

McGovern, J.H. 1981: The economics of extinction in Norse Greenland. In Wigley, T.M.L., Ingram, M.J. and Farmer, G. (eds.), *Climate and history* Cambridge: Cambridge University Press. 404–30.

McIlwaine, S.J. 1994: The influence of climate on the wave resource at the Islay Shoreline Wave Energy Demonstration Unit. *Weather* 49 no. 2, 65–71.

McLaren, A.S., Walsh, J.E., Bourke, R.H., Weavers, R.L. and Wittmann, W. 1992: Variability of sea ice thickness over the North Pole 1977–90. *Nature* 358, 224–6.

Maley, J. 1977: Palaeoclimate of the central Sahara during the early Holocene. *Nature* 269, 572–7.

Manley, G. 1974: Central England temperatures: monthly means 1659 to 1973. *Quaterly Journal of the Royal Meteorological Society* 100, 389–405.

Marsh, T.J. and Monkhouse, R.A. 1993: Drought in the UK. *Weather* 48, 15–22.

Matthews, J.A. 1975: Experiments on the reproducibility and reliability of lichenometric data, Storbreen gletschervorfeld, Jotunheim, Norway. *Norsk Geografisk Tidsskrift* 29, 97–109.

Matthews, J.A. and Shakesby, R.A. 1984: The status of the Little Ice Age in southern Norway: relative dating of neoglacial moraines. *Boreas* 13, 333–46.

Mercer, J.H. 1978: West Antarctic ice sheet and CO_2 greenhouse effect: a threat of disaster. *Nature* 271, 321–5.

Minami, K. and Meve H-V. 1994: Rice paddies as a methane source. *Climatic Change* 27, 13–26.

Mintzer, I.M. (ed.) 1992: *Confronting climate change* Cambridge: Cambridge University Press.

Monserud, R.A., Tchebakova, N.M. and Leemans, R. 1993: Global vegetation change predicted by the modified Budyko model. *Climatic Change* 25, 59–83.

Mount, T.P. 1994: Climate change and agriculture: a perspective on priorities for economic policy. *Climatic Change* 27, 121–38.

Nasralla, H.A. and Balling, R.C. 1993: Spatial and temporal analysis of Middle Eastern temperature changes. *Climatic Change* 25, 153–61.

Neumen, J. 1993: Climatic changes in Europe and the Near East in the second millennium BC. *Climatic Change* 23, 231–45.

Newton, W. and Fairbridge, D.W. 1986: The management of sea level rise. *Nature* 320, 319–21.

Nicholson, S.E. 1978: Climatic variations in the Sahel and other African regions during the past five centuries. *Journal of Arid Environments* 1, 3–24.

Nilsson, A. 1992: *Greenhouse Earth* Chichester: Wiley.

Oechel, W.C. et al. 1994: Transient nature of CO_2 fertilization in arctic tundra. *Nature* 371, 500–3.

Oerlemans, J. 1993a: Evaluating the role of climate cooling in iceberg production and the Heinrich events. *Nature* 364, 783–6.

Oerlemans, J. 1993b: Possible changes in the mass balance of the Greenland and Antarctic ice sheets and their effects on sea level. In Warrick, R., Barrow, E.M., and Wigley, T.M. (eds.), *Climate and sea level change* Cambridge: Cambridge University Press. 144–61.

Ogilvie, A.E.J. 1984: The past climate and sea ice record from Iceland. Part I: to AD1780. *Climatic Change* 6, 131–52.

Oliver, J. and Kingston, J.A. 1970: The usefulness of ships' logs in the synoptic analysis of past climates. *Weather* 25, 520–8.

Paren, J.G., Doake, C.S.M. and Peel, D.A. 1993: The Antarctic Peninsula contribution to future sea level rise. In Warrick, R.A., Barrow, E.M. and Wigley, T.M.L. (eds.), *Climate and sea level change* Cambridge: Cambridge University Press. 162–8.

Parry, M.L. 1976: The abandonment of upland settlement in southern Scotland. *Scottish Geographical Magazine* 92, 50–60.

Parry, M.L. 1977: *Climatic change, agriculture and settlement* Folkestone: Dawson.

Parry, M.L. 1990: *Climate change and world agriculture* London: Earthscan.

Parry, M.L., Carter, T.I. and Konijn, N.T. 1988: *The impact of climatic variations on agriculture. Vol 1. Assessment in cool temperate and cold regions* Dordrecht: Kluwer.

Pastor, J. 1993: The northward march of spruce. *Nature* 361, 208–9.

Peel, D.A. 1993: Cold answers to hot issues. *Nature* 363, 403–4.

Pethick, J. 1991: Mangroves, marshes and sea level rise. *Geography* 76, 79–81.

Pfister, C. 1981: An analysis of the Little Ice Age climate in Switzerland and its consequences for agricultural production. In Wigley, T.M.L., Ingram, M.J. and Farmer, G. (eds.), *Climate and history* Cambridge: Cambridge University Press. 214–48.

Pfister, C. 1985: Snow cover, snow lines and glaciers in central Europe since the sixteenth century. In Tooley, M.J. and Sheail, G.M. (eds.), *The climatic scene* 154–72.

Philander, S.G. 1990: *El Niño, La Nina and the Southern Oscillation* London: Academic Press.

Pincus, R. and Baker, M.D. 1994: Effect of precipitation on the albedo susceptibility of clouds in the marine boundary layer. *Nature* 372, 250–2.

Pittock, A.B. 1983: Recent climatic changes in Australia: implications for a carbon dioxide warmed world. *Climatic Change* 5, 321–40.

Pittock, A.B. 1994: Climate and food supply. *Nature* 371, 25.

Pittock, A.B. and Flather, P.A. 1993: Severe tropical storms and possible environmental effects. In Warrick, R.A., Barrow, E.M. and Wigley, T.M.L. (eds.), *Climate and sea level change* Cambridge: Cambridge University Press. 292–4.

Porter, S.C. 1981: Glaciological evidence of Holocene climatic change. In Wigley, T.M.L., Ingram, M.A. and Farmer, G. (eds.), *Climate and history* Cambridge: Cambridge University Press. 82–111.

Post, J.D. 1977: *The last great subsistence crisis in the western world*. Baltimore: John Hopkins University Press.

Pugh, D.T. 1993: Improving sea level data. In Warrick, R, Barrow, E.M. and Wigley, T.M. (eds.), 1993: *Climate and sea level changes: observations, projections and implications* Cambridge: Cambridge University Press. 57–71.

Rahmstorf, S. 1994: Rapid climate transitions in a coupled ocean-atmosphere model. *Nature* 372, 82–5.

Rampino, M.R. and Self, S. 1992: Volcanic winter and accelerated glaciation following the Toba super eruption. *Nature* 259, 50–2.

Rannie, W.F. 1983: Break-up and freeze up of the Red River at Winnipeg in the nineteenth century and some climatic implications. *Climatic Change* 5, 283–9.

Raper, S.C.B. 1993: Observational data on the relationship between climatic change and the frequency and magnitude of severe tropical storms. In Warrick, R, Barrow, E.M. and Wigley, T.M. (eds.), *Climate and sea level change* Cambridge: Cambridge University Press. 192–214.

Read, P. 1994: *Responding to global warming*. London: Zed Books.

Reilly, J. 1994: Crops and climatic change. *Nature* 367, 118.

Roberts, N. 1989: *The Holocene: an environmental history* Oxford: Blackwell.

Roberts, N. (ed.) 1994: *The changing global environment* Oxford: Blackwell.

Roberts, N., Taleb, M., Barker, P., Damnati, B., Icole, M. and Williamson, D. 1993: The timing of the Younger Dryas event in East Africa from lake level changes. *Nature* 366, 146–8.

Rosenberg, J. 1993. The MINK methodology: background and baselines. *Climatic Change* 24, 7–22.

Rosenzweig, C. 1994: Maize suffers a sea change. *Nature* 370, 175–6.

Rosenzweig, C. and Parry, M.L. 1994: Potential impact of climate change on world food supply. *Nature* 367, 133–8.

Rosman, K.J.R., Chisholm, W., Boutron, C.F., Candelone, J.P. and Gorlach, V. 1993: Isotopic evidence for the source of lead in Greenland snows since the 1960s. *Nature* 362, 333–5.

Royle, S., Robinson, J. and McCrea, A. 1994: Renewable energy in Northern Ireland. *Geography* 79, 232–45.

Sahagian, D.L., Schwartz, F.W. and Jacobs, D.K. 1994: Direct anthropogenic contributions to sea level rise. *Nature* 367, 54–7.

Salafsky, N. 1994: Drought in the rain forest: effects of the 1991 El Niño-Southern Oscillation event on a rural economy in west Kalimantani, Indonesia. *Climatic Change* 27, 373–90.

Sarmiento, J.L. 1993: CO_2 stalled. *Nature* 365, 697–8.

Schneider, S.H. 1990: *Global warming* Cambridge: Lutterworth Press.

Schonwiese, C-P. et al. 1994: Solar signals in global climatic change. *Climatic Change* 27, 259–81.

Siegenthaler, U. and Sarmiento, J.L. 1993: Atmospheric CO_2 and the ocean. *Nature* 365, 119–25.

Smith, D.I. 1993: Greenhouse climatic change and flood damages: the implications. *Climatic Change* 25, 319–33.

Stephens, G.L. 1994: Dirty clouds and global cooling. *Nature* 370, 420–1.

Stine, S. 1994: Extreme and persistent drought in California and Patagonia during mediaeval time. *Nature* 369, 546–9.

Street, F.A. and Grove, J.A. 1979: Global maps of lake level fluctuations since 30,000 years BP. *Quaternary Research* 12, 83–118.

Street-Perrott, F.A. 1993: Ancient tropical methane. *Nature* 366, 411–2.

Street-Perrott, F.A. 1994: Drowned trees record dry spells. *Nature* 369, 518.

Street-Perrott, F.A. and Harrison, S.P. 1985: Lake levels and climate reconstruction. In Hecht, A.D. (ed.), *Paleoclimate analysis and modelling* Chichester: Wiley 291–340.

Subak, S., Raskin, P. and Von Hippel, D. National greenhouse gas accounts: current anthropogenic sources and sinks. *Climatic Change* 25, 1993, 15–38.

Taylor, K.C., Hammer, C.U., Allen, R.B., Clausen, H.B., Dahl-Jenson, D., Gow, A.J., Gunderstrup, N.S., Kipfstuhl, J., Moore, J.C. and Waddington, E.D. 1993: Electrical conductivity measurements from the GISP2 and GRIP Greenland ice cores. *Nature* 366, 549–52.

Taylor, K.E. and Penner, J.E. 1994: Response of the climate system to atmospheric aerosols and greenhouse gases. *Nature* 369, 734–7.

Thomas, R.H., Sanderson, T.J.D. and Rose, K.E. 1979: Effects of a climatic warming on the West Antarctic ice sheet. *Nature* 277, 355–8.

Thompson, L.G. 1992: Ice core evidence from Peru and China. In Bradley, R.S. and Jones, P.D. (eds.), *Climate since 1500* London: Routledge. 549–71.

Thouveny, J., de Beaulieu, J.L., Bonifay, E., Creer, K.M., Guiot, J., Icole, M., Johnsen, S., Jousel, J., Reille, M., Williams, T. and Williamson, D. 1994: Climatic variations in Europe over the past 140 kyr deduced from rock magnetism. *Nature* 371, 503–6.

Titow, J.Z. 1960: Evidence of weather in the account rolls of the bishopric of Winchester 1209–1350. *Economic History Review* 12, 360–407.

Tooley, M.J. and Jelgersma, S. 1992: *Impacts of sea level rise on European coastal lowlands* Oxford: Blackwell.

Torn, M.S. and Fried, J.S. 1992: Predicting the impacts of global warming on wildland fire. *Climatic Change* 21, 257–74.

Turner, J. 1981: The Iron Age. In Simmons, I. and Tooley, M. (eds.), *The environment in British prehistory* London: Duckworth, 250–1.

Van Engelenburg, B. and Blok, K. 1993: Disposal of carbon dioxide in permeable underground layers: a feasible option? *Climatic Change* 23, 55–68.

Walker, G. 1993: Stopping an ice stream. *Nature* 365, 609–11.

Walsh, J.K. 1993: The elusive arctic warming. *Nature* 361, 300–1.

Warrick, R.A. and Barrow, E.M. 1991: Climatic change scenarios for the UK. *Transactions of the Institute of British Geographers* NS 16, 387–400.

Warrick, R.A., Barrow, E.M. and Wigley, T.M. (eds.) 1993: *Climate and sea level change* Cambridge. Cambridge University Press.

Warrick, R.A. and Farmer, G. 1990: The greenhouse effect, climatic change and rising sea level: implications for development. *Transactions of the Institute of British Geographers* NS 15, 5–20.

Weaver, A.J. 1993: The oceans and global warming. *Nature* 364, 192–3.

Weaver, A.J. and Hughes, J.M. 1994: Rapid interglacial climate fluctuations driven by North Atlantic ocean circulation. *Nature* 367, 447–50.

Webb, R.S. and Overpeck, J.T. 1993: Carbon reserves released? *Nature* 361, 497–8.

White, J.W.C. 1993: Climate change: don't touch that dial. *Nature* 364, 186.

Whiting, G.J. and Chantonn, J.P. 1993: Primary production control of methane emission from wetlands. *Nature* 364, 794–5.

Wigley, T.M.L. 1994: Climate change: outlook becoming hazier. *Nature* 369, 709–10.

Wigley, T.M.L, Ingram, M.J. and Farmer, G. (eds.) 1981: *Climate and history* Cambridge: Cambridge University Press.

Woodroffe, C. 1994: Sea level. *Progress in Physical Geography* 18, 436–51.

Woodward, F. (ed.) 1992: *Global climatic changes: the ecological consequences* London: Academic Press.

Wuebbles, D.J. and Edmonds, J. 1991: *Primer on greenhouse gases* Chelsea, Michigan: Lewis.

Wyman, R.L. (ed.) 1991: *Global climate change and life on Earth* London: Routledge.

Yarnal, B. 1985: Extra-tropical teleconnections with El Niño/Southern Oscillation. *Progress in Physical Geography* 9, 315–52.

Yim, W. 1993: Future sea level rise in Hong Kong and possible environmental effects. In Warrick, R.A., Barrow, E.M. and Wigley, T.M.L. (eds.), *Climate and sea level change* Cambridge: Cambridge University Press. 349–78.

Zahn, R. 1994: Core correlations. *Nature* 371, 289–90.

Index